P9-DWC-439
3 2052 00515 1199

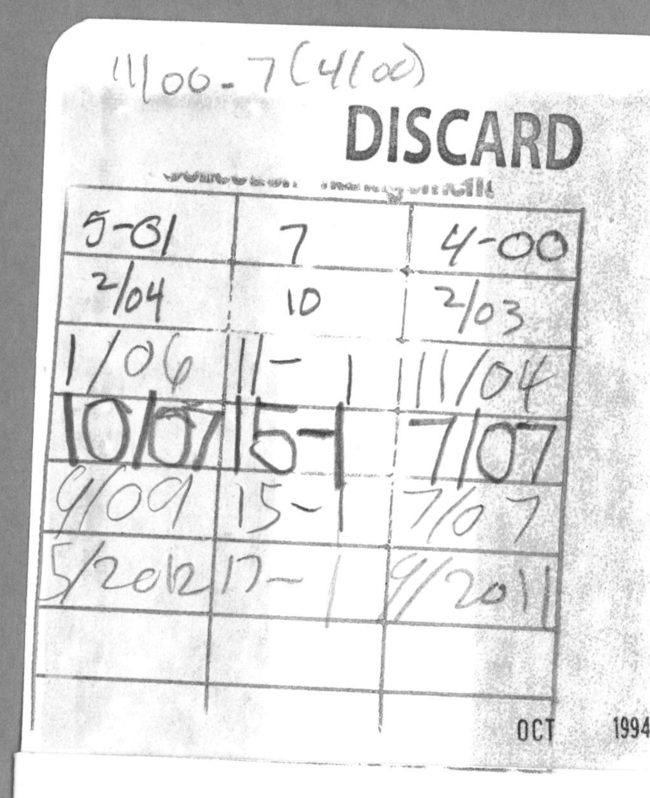
11/00-7 (4/00)
DISCARD
5-01 7 4-00
2/04 10 2/03
1/06 11-1 11/04
10/07 15-1 7/07
9/09 15-1 7/07
5/2012 17-1 9/2011
OCT 1994

Future Harvest

Volume 5 in the series
Our Sustainable Future

Jim Bender

Future Harvest

Pesticide-Free Farming

University of Nebraska Press
Lincoln and London

Manufactured in the United States of America

The paper in this book
meets the minimum requirements of
American National Standard for
Information Sciences – Permanence of Paper for
Printed Library Materials,
ANSI Z39.48–1984

Library of Congress Cataloging-in-Publication Data

Bender, Jim, 1950–
Future harvest : pesticide-free farming / Jim Bender.
p. cm. — (Our sustainable future; v. 5)
Includes bibliographical references and index.
ISBN 0-8032-1233-X (alk. paper)
1. Organic farming—Middle West. 2. Organic farming—Nebraska—Case studies.
3. Sustainable agriculture—Middle West. 4. Sustainable agriculture—Nebraska—Case studies.
I. Title. II. Title: Pesticide-free farming. III. Series.
S605.5.B46 1994
630'.977—dc20
93-15807
CIP

• • •

This book is dedicated to
Edward Bender
and
Patricia Williams
who know what it means,
and *what it meant*

• • •

Contents

• • •

Illustrations

Figures

Maps

Make everything as simple as possible, but not more so.

—commonly attributed to Albert Einstein

• • •

Preface

It would seem that farming should be able to take care of itself. Several million farmers live on the land, presumably coming to an increasingly more sophisticated understanding of agriculture and of the nuances of their own farms. Their level of personal satisfaction must be high, for large numbers of them hope that their sons and daughters will carry on the farming tradition. Thus, there should be abundant incentive to farm for the future as well as for the present.

However, something has gone profoundly wrong. Instead of conserving and improving our agricultural resources, modern farming is broadly implicated in their degradation. Much of our topsoil is gone. Fragile lands are devastated. Ground water is polluted with insecticide, herbicide, and fertilizer residues. Aquifers are being depleted. Pesticide residues are found in some foods.

Yet coexisting with this kind of self-destructive farming is another kind of agriculture, one which avoids or greatly reduces such problems. It is referred to variously as "alternative agriculture," "stewardship-oriented agriculture," "sustainable agriculture," "low-input agriculture," or "regenerative agriculture." Although differing somewhat in specific configurations, this approach is generally characterized by broadly diversified cropping, with crops and livestock integrated into the same operation, and to a large extent by the use of alternatives to pesticides and synthetic fertilizers.

While interest in this kind of farming is growing and it is beginning to receive institutional support, most conventional farmers are reluctant to commit to it. There are many reasons for this hesitancy, some of which I describe in chapter 1. However, my purpose in writing this book is not to dwell on the reasons why conventional farming continues. Rather, I provide an accessible, step-by-step model of conversion to stewardship-oriented agriculture, and I offer a rationale for pursuing pesticide-free farming. Using my experience in making the transition on a medium-sized farm in southeastern Nebraska, I discuss in considerable detail how to convert to an alternative system in the Midwest. Many of the techniques and strategies I describe are widely applicable well beyond the Midwest, and even when the specifics do not apply directly, the general principles illustrated are appropriate to a variety of farming situations.

My relationship with stewardship-oriented agriculture began in 1975, when I returned to Nebraska from graduate studies in the humanities and found myself with an opportunity to participate in the operation of a four-generation family farm. Because I was a committed environmentalist, I became increasingly fascinated with the idea of stewardship agriculture. It seemed to me that the only possible objection to it was that it would not work. Therefore, I decided that a useful way to express my environmental convictions would be to try to create a farming operation without the environmental problems associated with conventional farming.

The challenge was twofold. The first was to fight soil erosion in every way possible in the context of crop production. In this I continued a long-standing family commitment. The special challenge was that according to the U.S. Soil Conservation Service, southeastern Nebraska suffers from some of the worst soil loss in the country (U.S. Soil Conservation Service 1989). Most of the land I farm has been categorized as potentially highly erodible, a situation requiring special procedures to minimize soil loss.

The second goal was to farm without herbicides, insecticides, fungicides, or synthetic fertilizers. And, of course, to farm well, for otherwise the project would have little point. This proved to be more difficult to achieve than I had initially thought. At first my determination vastly exceeded my knowledge, with rather comical results. So I backtracked and planned a schedule of transition to correspond with my developing skills and experience. I felt that the overall setting would provide an appropriately rigorous test of viability. The farm size (642 acres), the location on the western edge of the dryland Corn Belt, and the need to integrate conservation and organic practices in one system would provide a variety of challenges.

My struggle to be free of herbicides and insecticides continued until 1980.[1] At the time, this was a very solitary project. It proceeded in spite of a government farm program that discouraged it, without assistance from the extension service or trade magazines, and with little positive interest from neighbors.

When I needed experimental results—which was often the case—I had to experiment myself. Thus, an inspiration for this book is to help others avoid the needless frustrations and false starts associated with conversion.

One of the central subjects of this book is conversion to and organization of chemical-free farming. I will concentrate much more on eliminating pesticides than on eliminating synthetic fertilizers. The reason is that my longer experience with pesticide-free farming (13 years) gives me confidence to make many specific recommendations. Furthermore, much of what is required to eliminate pesticides is identical with what is necessary to position a farm for alternative sources of fertility. The fertilizer goal requires not different procedures, but only a few additional ones.

In chapter 1, I explore the context which contemporary agriculture creates for the possibility of change as well as preliminary reasons why pesticide-free farming should be taken seriously as a goal. Chapters 2 and 3 deal, in a practical and detailed way, with problems and issues of conversion. Chapter 4 examines the role of livestock in alternative systems, blending both practical topics and policy issues. Chapter 5 compares conventional and alternative agriculture from several perspectives. And in chapter 6, I address four popular arguments used against alternative agriculture and in support of conventional farming. In these later chapters, I generally use organic farming to represent alternative agriculture. Answering the criticisms of organic agriculture—the most restrictive in terms of inputs and, hence, the most implausible from the point of view of conventional farming—makes the strongest possible case for alternative agriculture in general.

Because this book is based on my experience farming where the dryland Corn Belt and the Great Plains meet, there are many aspects of pesticide-free farming it does not address. For example, it is silent on organic vegetable farming or orchards. Those stories will be told by the men and women doing and studying those kinds of farming. Furthermore, for every procedure I describe, there are probably other nonchemical methods being used by other farmers that are at least as effective. And obviously some of the detail is provisional and open to improvement. Nevertheless, the general system which I present of organizing a farm and of solving farming problems is, I believe, the appropriate alternative to the chemical-intensive agriculture predominant in wide regions of this continent and, indeed, in much of the world. With each passing year, my confidence in the viability of the general structure of alternative agriculture only increases.

For me, one of the sweetest moments of my life came in 1980, when I eliminated all pesticide use on my field crops, thereby successfully matching my practice with my convictions. Since then, my satisfaction in doing this kind of farming—participating in an organic biological process, lessening dependence on purchased inputs, preserving resource and environmental quality, meeting stimulating challenges—has continued and grown. I hope that this book conveys not only the feasibility of pesticide-free farming but also the joy it brings.

Acknowledgments

Portions of this book have previously been published in different form: part of Chapter 1 as "Agricultural addiction," in *American Journal of Alternative Agriculture* 5, no. 4 (1990): 168–69; parts of Chapter 2 as "Converting to pesticide-free farming—coping with institutions" in *Journal of Soil and Water Conservation* 45, no. 1 (January–February 1990): 96–98, and "How non-operator farmland owners can promote land stewardship," in *American Journal of Alternative Agriculture* 2, no. 3 (1987): 98; parts of Chapter 3 as "Whip soybean weeds now," in *The New Farm* (May–June 1985): 28, "Get the most from your rotary hoe," in *The New Farm* (March–April 1986): 23, and "Save soil, control weeds—without chemicals," in *The New Farm* (March–April 1989): 17; part of Chapter 4 as "Does organic farming require too much livestock?" in *American Journal of Alternative Agriculture* 3, no. 1 (1988): 2, 40; and parts of Chapter 5 as "Substituting management for cash inputs in organic systems," in *American Journal of Alternative Agriculture* 4, no. 1 (1989): 2, 46. I gratefully acknowledge permission from the editors of *The Journal of Soil and Water Conservation, American Journal of Alternative Agriculture,* and *The New Farm* to reprint the material here in revised form.

Many individuals have contributed to this project in myriad ways. They include neighbors, friends, and the many young persons who have worked on the farm. Several people deserve special mention. My late mother, Virginia Bender, in her role as a farm owner, supported our goals even in 1975 when such matters seemed more improbable. My father, Edward Bender, has spent much of his retirement years engaged in our project in every imaginable way. His wisdom about soil conservation, and his enthusiasm, have been especially invaluable. Patricia Williams's participation has also been at virtually every level, particularly animal and tree care. She has been the primary editor of everything I have written about agriculture. In response to such loving dedication by them I must simply say thank you.

For many years Charles Shapiro has elegantly combined his role as a friend to encourage, and his role as an agricultural scientist to suggest, criticize, and provide perspective.

Only I, however, am responsible for what remains undone or poorly conceived.

1

• • •

Change: Goals and Obstacles

What is characteristic of conventional agricultural systems in the Midwest? Among their key features are the separation of livestock from crop production, minimal crop diversity, and the lack of long, multiyear crop rotation. And, of course, central to them is the routine use of pesticides and synthetic fertilizers. In the Midwest, conventional systems are typically limited to one or two cash grain crops planted in late spring, usually corn and soybeans. Nevertheless, not all agricultural systems fall neatly into the conventional or pesticide-free categories. Those systems that modify the conventional format with livestock or legumes or other cropping diversity or which substantially reduce chemical inputs are less incompatible with the agriculture that I will be describing and defending.

In 1980 mine was just one of several alternative farming systems being developed across the country. They were arising within a rather indifferent agricultural community at that time. The idea that the main obstacle to alternative farming is simply the perception that it will not work has not held up. It was necessary to search for a broader context in which to understand the problem. This context includes several components. One is the federal farm program which, among other things, has penalized crop rotations. Another is that agricultural research has largely ignored organic methods. A third is several major trends in contemporary agriculture that have been caused by or have resulted in dependence upon agricultural chemicals. Yet another component is that organic farming has been misunderstood in a way that understates its advantages and overstates its disadvantages. The first two components—the federal farm program and agricultural research—have been discussed at length in the recent National

Research Council's study of alternative agriculture (1989) and by Francis, Flora, and King (1990, chapter 13). I will, therefore, focus on the last two.

Several recent trends in American agriculture have heightened farmers' dependency upon agricultural chemicals. One trend is the separation of livestock from farms. In beef cattle production, for example, both small feeding enterprises and cow-calf operations are disappearing from farms. In just the five years between 1975 and 1980 there was a decline of 10% in cow-calf operations on individual farms; among small herds—those with 20 to 99 cows—the loss was 14% (Van Arsdall and Nelson 1983, 37). But that is only part of the problem, for in the process, the costly infrastructure of livestock production—fences, watering systems, barns, handling facilities, and so forth—is being destroyed.

A second trend in contemporary agriculture is increasing farm size and decreasing number of farms. Probably any size farm can be organic, but as organic farm units become larger, management challenges increase greatly. The competence and dedication of the hired help becomes crucial.

A third trend is the separation of land ownership from farming. Over half of the farmland in the United States is farmed by persons who do not own it (Lewis 1978, 7). Farm tenants who wish to convert to stewardship agriculture face two serious problems. First, they are adopting methods of operation that pay off primarily in the long term. Second, they are replacing many cash expenses usually shared by the landowner with management responsibilities that are largely their own burden.

Still another modern trend is the generational distance of today's farmers from nonchemical farming. Many young farmers are now two generations away from organic farming procedures. Their parents have also known only chemical agriculture. The learning which occurs when parents show their children how to do something is special. It cannot be easily compensated for by extension-service assistance, formal education, or other impersonal sources of information.

A fifth trend is the great increase in part-time farming. Combining farming with a job off the farm has become increasingly popular. It is easy to see how, in the press for time to do the farm work, chemicals can be especially enticing.

And, finally, a significant factor is the difficulty of dealing with certain perennial weeds which, although they can be suppressed with herbicides, continue to spread. This means that a conventional farmer can subdue the weeds sufficiently to grow and harvest the crops, but the basic infestation problem actually gets worse. The reason is their capacity to spread asexually by roots and general tillage. Examples of such weeds are field bindweed (*Convulvus ar-*

vensis L.), Canada thistle (*Cirsium arrense* L.), and hemp dogbane (*Apocynum cannabiuum* L.).[1] Herbicides will mask these weed problems, but once applications stop, the problem can be virtually overwhelming. This is almost a classic pathology of dependency.

Each of these trends contributes to the dependency upon agricultural chemicals, making it difficult for farmers to evaluate contrasting agricultural practices objectively. Furthermore, the trends are intensifying. Thus they make the conversion process a vastly greater challenge, and they seriously complicate the development of a context for change. Moreover, there is a tendency for the defenders of conventional agriculture to understate the strengths and overstate the weaknesses of organic and other alternative systems. This distortion has created at least the following three myths about alternative agriculture.[2]

The first myth is that the main or only motivation for and benefit from alternative systems is satisfying a personal ethical or environmental principle. However, it is much more than this. These systems effectively address such agronomic problems as weed and insect cycles, fertility management, soil compaction, risk management, and nutrient recycling. By the early 1980s, I realized that I was developing a system that was yielding much more than mere consistency with my environmental and ethical convictions.

Another myth is that alternative systems are simply a reversion to an earlier, more primitive era of agriculture, such as that practiced in the early 1900s. That is false for several reasons. For one thing, the economics of contemporary farming require much greater productivity. Good organic farmers do not accept the low yields of earlier times. They use management strategies and production techniques far removed from those of their ancestors. Second, the cropping patterns of today are significantly different from those of the early 1900s. For example, soybean production has increased from 5 million bushels in 1924 (Martin, Leonard, and Stamp 1976, 691) to approximately 2 billion bushels in the 1980s. And although soybeans present special weed control problems, many farmers accept the challenge of growing them—without the use of herbicides. A final difference between modern alternative agriculture and the farming of earlier years is that our ancestors were quite unsophisticated about soil conservation, often with disastrous consequences. Today's alternative farmers accept the considerable challenge of combining organic agriculture *and* soil conservation into one system.

The third myth that afflicts pesticide-free agriculture is that only conventional systems are convenient. Granted, it is not easy to measure comparative convenience. And it is true that certain components of alternative systems—for example, livestock—can hardly be viewed as convenient. However, as a farmer comes to the final stages of conversion, a very different sense of the situation can emerge on well-organized farms. Because of the interrelatedness of the components of pesticide-free systems, often the actions taken to meet one objective also contribute to attaining others in ways that are just the opposite of what occurs in conventional systems. This is one reason why pesticide-free farmers often claim that their farming becomes easier each year.

Thus it is clear that trends in contemporary agriculture and myths surrounding pesticide-free farming work together to create a context which is not conducive to change. Add to this that federal farm programs penalize crop rotation and that agricultural research has largely ignored organic methods, and it becomes easy to understand why for many farmers the goal of farming without chemicals seems odd and impossible.

Goals: Reducing Versus Eliminating Pesticides

In the discussion of the conversion process, a central component will be *gradualism*. Gradualism refers to, among other things, a slow, measured, but steady reduction of chemical use. That, however, pertains to conversion, moving from heavy reliance on agricultural chemicals to something else. It should not be confused with goals. What a farmer's ultimate agricultural goals might be is another subject.

As the last decade of this century begins, there is an emerging consensus that agriculture has environmental problems. It is even stated or implied in television advertisements for pesticides. In response, a popular objective is to somewhat *reduce* pesticide usage. It refers to such practices as more carefully calibrating sprayer nozzles, using pesticides on an "as needed" basis rather than as insurance, and banding rather than broadcasting herbicides. As a strategy, it is even being dubbed "sustainable" (Buttel and Gillespie 1988). The vagueness of these terms makes possible a variety of political, economic, and research objectives. Also, for those who continue to suspect that in pesticide-free and organic farming there is something fanatical, something extra-agronomic, the reduction objective is probably reassuring.[3]

There is another component to this subject. Some opponents of alternative agriculture would leave proponents of pesticide-free farming in a conceptual squeeze. If the proponents defend or practice pesticide-free agriculture, the charge of fanaticism or irrational inflexibility arises. If the proponents defend instead a position that permits minimal pesticide usage in some cases, the position is characterized as not that much different from best management goals of mainstream agriculture. The subject has, therefore, been conceived in such a way that proponents of alternative agriculture would be left with no plausible position.

As the reader begins to make decisions, I offer preliminary defense of the goal of complete freedom from pesticides—without fanaticism, without arbitrariness. A broader rationale emerges in chapters to follow.

It would not be wise to make a case for pesticide-free farming by asserting that any use of pesticides, no matter how small, could leave us in peril. That is not a defensible position. The argument has to do instead with what will be required to break the addictive grip of pesticide usage upon agriculture. That consideration, along with various environmental concerns, is the basis for going beyond merely reducing pesticide usage to a different kind of agriculture.

Of the numerous environmental issues swirling around pesticides, I will mention just one. Our understanding of ground water contamination by pesticides evolved rapidly in the 1980s. An early view was that contamination is the result of accidents and other unusual events. That view was quickly repudiated. In 1988 the Environmental Protection Agency's Office of Pesticide Programs released a report entitled *Pesticides and Ground Water Data Base: 1988 Interim Report* (Williams et al. 1988). This study not only acknowledged but also sought to document the extent of ground water contamination by normal agricultural practices. It identified 46 pesticides in 26 states leaching into ground water as a result of normal use (3–11).[4] The implications for an environmentally concerned farmer are obvious. If the conclusions of the interim report and other investigations of this subject are correct, then the pesticide "reduction" objective has not been demonstrated to be an appropriate solution to the problem.

This book will emphatically not be concerned with tinkering with conventional systems. A central strategy will be to describe a contrasting system that, among other things, steadily reduces and finally eliminates the incentive to use pesticides. Consider the other approach, which is to reduce in some way chemical usage in a conventional system. Since conventional systems have been made possible by the emergence of agricultural chemicals, it should hardly be surprising that these systems will continue to generate incentives—not to say pressures—to continue using pesticides. Consider, for example, a farmer with a conventional system on 640 acres with corn and soybeans in rotation. In such a system, weed control, whatever form it takes, must occur on all the acres in crops in a two- to three-week period. That puts extraordinary pressure upon the farmer to incorporate some preplant herbicide as insurance and at least to be prepared to use postemergence herbicide as a bailout. The point is that whatever a conventional farmer's environmental dedication, he or she labors within a system that exerts continual and considerable pressure to use pesticides.

Athletic coaches and other motivators know that appropriate goal selection is important to achievements. An unrealistic goal loses its motivating force, and overly modest goals can circumscribe one's accomplishments. Here there are countless applications to making changes in farming practices. Consider again the example of the conventional farmer given above. Suppose that he or she wants to reduce herbicide usage in row crops but continues to retain the post-emergence herbicide bailout option. Doing so reduces the incentive to gain proficiency in strategies for avoiding postemergence herbicides, strategies such as those discussed in chapter 3. I am suggesting that the goal of being pesticide-free can by itself push a farmer to more and useful skills and knowledge than he or she would otherwise attain.

How has the "reduction" goal worked out so far? The specter of self-deception lurks over subjects of coping with dependencies. Virtually all conventional farmers with whom I visit assert that they are reducing pesticide usage. Perhaps the emerging tendency for pesticides to be sold and applied in more concentrated mixtures contributes to this perception. However, greater concentration does not imply less active ingredient or expense, and pesticides are being used on more and more acres. For example, in 1990 in the 47 corn states, herbicides were used on 92.4% of acreage. Herbicides were applied to 94.8% of soybean acres in 29 states (U.S. Department of Agriculture 1991, 3; Osteen and Szmedra 1989, 13).[5]

In their U.S. Department of Agriculture publication entitled *Agricultural Pesticide Use Trends and Policy Issues,* Osteen and Szmedra (1989) write:

> The dramatic change in the emphasis of pesticide policy toward protection from health, safety, and environmental risks has had little effect on the percentage of acreage treated with pesticides. Acreage has steadily increased since 1972. Continued growth in herbicide use has been especially dramatic. However, cancellations, suspensions, and use restrictions have affected the individual compounds and the mix of pesticides used. (44)

Figures 1 and 2 show the high proportion of corn and soybean acres treated and the generally increasing trend over the years.

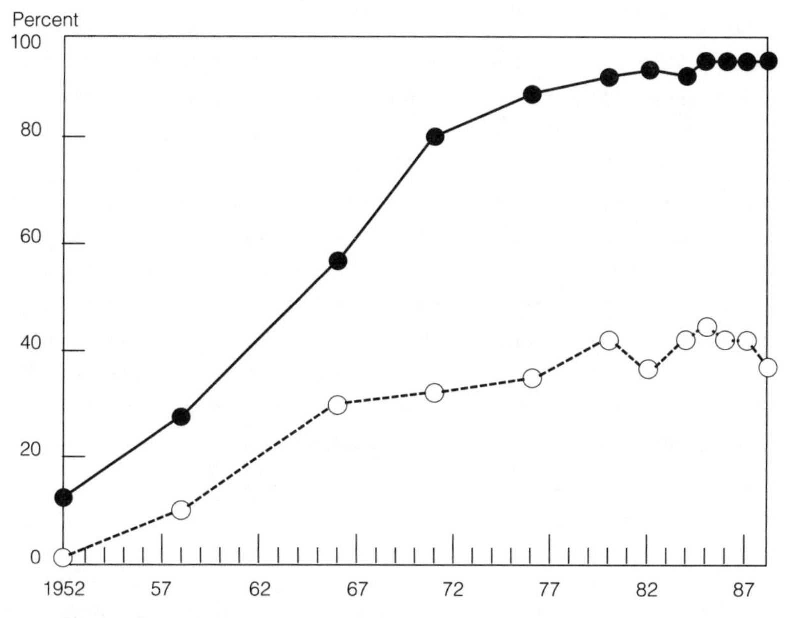

Figure 1. Share of corn acreage in the United States treated with pesticides. Redrawn from Osteen and Szmerdra 1989, 13.

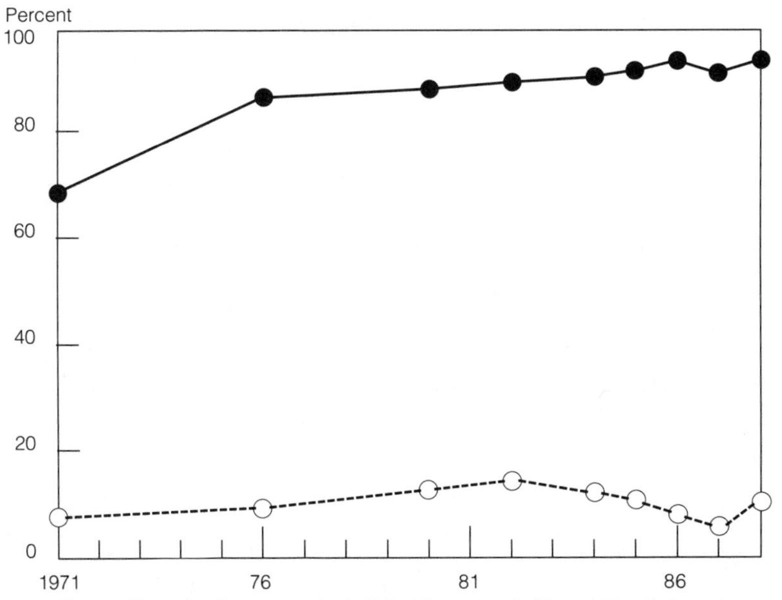

Figure 2. Share of soybean acreage in the United States treated with pesticides. Redrawn from Osteen and Szmerdra 1989,13.

I conclude that in general there is nothing in recent pesticide products, usage trends, or practices to indicate reduced reliance by conventional farmers.

Conclusion

That parts of the agricultural community would simultaneously call for "reduced" pesticide usage and also tolerate or promote trends such as no-till, separation of livestock and crop production, larger farm units, and declining crop rotation—all of which increase dependence upon pesticides—is indeed a curious situation. A starting point for understanding this paradox—and revising agricultural policy that is at odds with itself—would be a clearer sense of what pesticide-free agriculture entails and why there are concerns about pesticide trends. These subjects are included in the ensuing chapters.

2

• • •

Conversion

I have stopped using chemicals twice in my farming career. The first time was in 1975, when I took over the farming operation. The results of this attempt were unsatisfactory—even comical. I now understand the reasons for my early problems. Taken together, those reasons constitute a compelling argument for gradual, rather than immediate and complete, conversion. Among the conditions generally present at the outset are the following:

- The farmer seldom has enough knowledge and experience to deal with the many and various situations that will be present or with the problems that will arise.
- The machinery available will usually not be appropriate for the new style of farming.
- There will usually be herbicide carryover on the fields.
- A distorted weed situation may be present.
- Insect control may be dependent on pesticide use.
- Necessary infrastructure such as livestock facilities and soil conservation structures are often absent.
- Alternative sources of fertility are usually not developed and organized.

Any one of these situations by itself can seriously compromise conversion. For example, in my own case I did not sufficiently appreciate the crucial importance of early weed control in row crops. I didn't even have a rotary hoe or a tractor with appropriate ground speeds for hoeing. Another problem was that triazine-based herbicides, which were still popular at the time, caused us to lose the entire oats planting the first year and even carried over into the next. On one farm there was an acute hemp dogbane problem,[1] and the same farm was highly susceptible to greenbug damage.

Agronomics

Overly eager conversion simply produces problems, disappointments, and frustrations which provide little useful learning experience and which present numerous temptations to give up. I want to describe a conversion process which minimizes these difficulties. In it, all components—knowledge, skills, facilities, and processes—progress in a mutually reinforcing manner.[2]

Begin by considering steps which are necessary as background preparation for conversion and which can commence even while continuing to farm conventionally. I recommend the following: an ambitious soil conservation program, crop diversification and rotation, livestock production, and small-scale experimentation with pesticide-free methods.

SOIL CONSERVATION

In this book I have assumed very few things as self-evident, but one of them is an active commitment to soil conservation. Neglecting topsoil conservation affronts the very concepts of rationality and ethics. However, sometimes persons who accept basic soil conservation goals do not see how they apply to their farm or to a particular field. Accordingly, instead of justifying soil conservation, I will discuss the rationale for placing so much emphasis on it, especially the important role of terraces and waterways in minimizing soil loss and maximizing moisture retention.

Many soil conservation challenges converge in the Midwest. Much of the soil in our particular locale is designated by the Soil Conservation Service as highly erodible. Our weather is noted for strong winds and extremely intense thunderstorms of short duration that test the best conservation systems. We experienced many of these storms in the 1980s, especially during planting season, when freshly tilled soils are most vulnerable. These, of course, are challenges for all farms in our region. Consider that in that decade Weeping Water, Nebraska, counted eight May and June rainfalls of two inches or more, four of which measured from four to seven and a half inches (data from unpublished

"Record of River and Climatological Observations" sheets compiled by Sterling Farley Amick, U.S. National Weather Service Observer, Weeping Water, Nebraska).

In chapter 5, I discuss in detail what I consider to be a comprehensive approach to soil conservation. It does not set the unrealistic goal of zero soil erosion, but rather the realizable goal of doing everything possible to fight it in the context of crop production. Just what that entails is outlined in the farm profile presented in the appendix. Among other things, it rejects the idea that terraces will suffice in one field and high residue tillage in another. Insofar as possible, we want both, as well as other procedures, operative everywhere. So the reader agreeing with soil conservation goals, but judging that they do not apply (at least not very much) to his or her own situation, should visit with local Soil Conservation Service technicians before drawing such a conclusion. These experts will have considerable information about soil conservation challenges in your region and, perhaps, on your own farm. Even on flat land, water runoff and wind erosion must be managed.

The primacy of terraces. Because a key feature of this approach is the construction of terraces and waterways, I will present here a preliminary justification for their incorporation into the early stages of the conversion process. Our commitment to terraces (closely spaced) and waterways (over 30 miles of terraces and 25 acres of grassed waterways on 642 acres) is a much more aggressive conservation strategy than merely employing the currently popular conservation tillage. There are three primary reasons why they are important to a farmer preparing for pesticide-free farming.

One reason is that the diversified crop mix and the associated tillage patterns of a pesticide-free farm do not always result in enough crop residue remaining on the fields to effectively prevent erosion. For example, soybean residue is a notoriously insubstantial cover, one which winter grazing can reduce still more. What is left is often destroyed by the first disking in spring. Fields of winter wheat which have attained only moderate growth going into winter are susceptible to water erosion, especially before the soil is frost-free in the spring. Another cause of residue shortages is the removal of crops for silage. Finally, while moldboard plowing remains defensible for such specialty tillage situations as a field with perennial weed problems, it can virtually eliminate residue.

Table 1. Comparison of soil loss from terraced and unterraced watersheds	
Location and Length of Record	Watershed Characteristics (Terrace grades are 0.25% unless noted)
Guthrie, Oklahoma 1930–38	Unterraced; contoured
	Multiple, level, open end terraces; R=0.52†
Clarinda, Iowa 1934–41	Unterraced; corn/small grain rotation, farming parallel to field boundary
	Multiple terraces; contoured; corn/small grain rotation; grassed outlet channel
	Unterraced; contoured; corn/small grain/meadow rotation
	Unterraced; contoured; corn/small grain/meadow rotation
	Multiple terraces; contoured; corn/small grain/meadow rotation; grassed outlet channel
Bethany, Missouri 1935–42	Unterraced; contoured; grassed waterways
	Multiple terraces; contoured; grassed outlet channel
Tyler, Texas 1933–41	Unterraced; contoured
	Single terrace; contoured; soil loss measured at outlet of terrace
Zanesville, Ohio 1934–42	Unterraced; contoured
	Single terraced; contoured; soil loss measured at outlet of terrace
LaCrosse, Wisconsin 1933–36	Unterraced; contoured
	Single terrace; contoured; soil loss measured at outlet of terrace‡
1937–43	Unterraced; contoured; filter strip over lower 1/3
	Single terrace; contoured; soil loss measured at outlet of terrace‡

*Adjusted soil loss is measured soil loss divided by LS and P factor values to adjust sheet and rill erosion to a common base for terraced and unterraced watersheds.

†A watershed with a level terrace was the only one suitable for comparison. Sediment yield from a 0.25% grade terrace is 1/0.52 that from an open end, level terrace.

Slope Length, λ (ft.)	Slope Steepness, s (%)	USLE LS	Soil Loss Measured (tons/a)	Soil Loss Adjusted* (tons/a)
353	5.0	1.00	813.0	813.0
69	3.0	0.26	42.8	318.4
221	7.7	1.50	47.7	33.9
56	8.8	0.86	6.2	12.0
317	5.2	0.98	12.9	13.4
238	8.4	1.64	8.0	4.9
50	10.0	0.95	0.9	0.9
122	9.0	1.30	23.6	18.3
56	7.0	0.62	4.9	7.6
168	7.5	1.17	51.7	44.2
63	6.0	0.53	5.4	10.3
251	14.0	3.63	167.3	45.9
69	10.0	1.15	33.9	29.4
300	15.0	4.43	243.5	54.9
69	10.0	1.15	37.5	32.6
300	15.0	4.43	37.9	16.9
69	10.0	1.15	8.9	8.0

‡No uniform 0.25% grade terrace available. A variable 0 to 0.5% grade terrace was used. Data from other locations showed that the two give similar soil losses.

Source: Based on Foster and Highfill 1983, 50.

(For a comparison of soil loss from terraced and unterraced watersheds, see table 1.)

A second reason for early emphasis on terraces and waterways stems from requirements of the 1985 federal Food Security Act. Although our own soil conservation system was virtually complete before this legislation passed, it now is a reality for all farms with highly erodible land. The mandatory soil conservation plans which do not include terraces and waterways are more restrictive about rotations and tillage. Thus, permanent conservation structures help to preserve the cropping and tillage flexibility required for conversion.

A third reason for using terraces and waterways is to maximize the absorption of water. Their superiority over contouring was demonstrated in an eight-year study in Iowa. The location featured loess deposits with soils of the Marshall, Monona, Ida, and Napier series with slopes ranging from 2% to 18%. The researchers found that

> surface runoff was 10 percent of water yield from the terraced watershed and 65 percent of water yield from the contoured watersheds. Peak runoff rates from terraced-corn and grassed watersheds were 10 percent of those from contoured-corn watersheds. Sheet erosion averaged 24 tons per acre per year on the contoured-corn watersheds and less than 1 ton per year from the terraced-corn watersheds. Gully erosion averaged 450 tons per year from the contoured-corn watershed, but was negligible on the terraced and grassed watersheds (Spomer, Saxton, and Heinemann 1973, 168).

Results such as these impress upon us the fact that the considerable initial costs of terraces and waterways must be put into a long-term perspective.[3]

In summary, terraces and waterways are a key to combining soil conservation in difficult circumstances with pesticide-free farming. These structures maximally prepare a farm for various residue situations and weather extremes and for crop diversification. For all of these reasons, I urge that the preparation and implementation of the conversion process include as ambitious a conservation plan as possible. I will be concerned repeatedly in this book with explaining how nonchemical cropping and conservation can coexist and often complement each other in one operation. The federal Soil Conservation Service provides technical assistance at no charge, including help in identifying needs and establishing priorities. Major conservation work will, of necessity, have to be spread out over several years. However, this does not mean that parts of the farm must be allowed to suffer continuing erosion. Fields that cannot be scheduled for immediate work can be placed in government set-aside programs or planted to small grains or alfalfa to minimize erosion.

CROP DIVERSIFICATION AND ROTATION

A second important component of preparing for conversion is to develop and begin following a plan for crop diversification and rotation. This is indispensable to success, lying at the heart of many strategies, including those for weed and insect control as well as for achieving satisfactory crop yields. Three crops which I suggest considering seriously for early diversification are alfalfa, oats, and grain sorghum. The details of rotation will be postponed until the next chapter.

For me, alfalfa does more to make pesticide-free farming possible than any other crop. Its advantages and attributes read like a listing of the needs of pesticide-free farming:

- It is a significant source of nitrogen.
- It promotes tilth.
- It has a powerful weed-suppressing effect (technically called "allelopathy").
- It helps reduce the problem of soil compaction.
- It is a superior means of soil conservation.
- It is an excellent crop during drought.
- It is a constituent of most livestock operations.
- It spreads the workload.
- It is usually quite profitable as a cash crop.
- It can be processed in various ways.

Thus, it is easy to see why alfalfa is a superior first choice for diversification. It simultaneously addresses the concerns of weed control, fertility, and sustained profitability.[4]

A good choice for the second crop to introduce is oats.[5] Among its advantages are these:

- It provides good soil conservation.
- It is ideal for establishing legumes and grasses.
- It helps to suppress weeds.
- It is versatile, providing grain and straw which can be either used in the livestock operation or sold.
- Its production cost is low.

Another feature which makes oats attractive for the conversion process is that it provides a bailout option for special weed problems. If weeds overtake the oats in a difficult field or during a difficult year, the entire crop can be put up as hay.

A third crop to consider introducing early in the diversification process is grain sorghum.[6] Grain sorghum has the following useful characteristics:

- It uses nitrogen more efficiently than corn (Sander and Frank 1980).
- It has a strong weed-suppressing effect (Miller and Donahue 1990, 195).
- It performs comparatively well in drought (Martin, Leonard, and Stamp 1976).
- It somewhat interrupts insect cycles (for example, when used in rotation with corn).

However, it should be noted that sorghum is not as well suited to all regions of the midwest and plains states as alfalfa and oats. But for those areas where it can be grown, it is an excellent third choice.

Many other crops are suited to a system of chemical-free farming, including wheat and clover. The three I have discussed are simply excellent starting points for crop diversification.

LIVESTOCK

In addition to soil conservation measures and crops, another aspect of planning for and converting to an alternative system is incorporating livestock into the farming operation. The role of livestock is crucial. I am convinced that, in most cases, pesticide-free farming will not be viable without livestock. The general reasons for this are as follows:

- Livestock make it economically more feasible to develop appropriate soil conservation measures and to eliminate pesticides by making supportive crop rotation viable.
- Livestock provide financial stability.
- Livestock recycle nutrients.

In chapter 4, I elaborate on these points and discuss in detail the practical considerations for including livestock.

However, the obvious question that arises is: What kind of livestock? Ruminants are highly appropriate for the kind of system that I am describing, primarily because of their capacity to utilize forage efficiently. However, converting farmers who already have livestock should probably continue with what they have and know.

EXPERIMENTATION

The final component to incorporate in the early stages of conversion is small-scale experimentation with pesticide-free methods. The point is to have an early and direct learning experience. To understand why, think of learning a new game or computer program. After reading the instructions for so long, there comes a point when there is really no substitute for actually trying to play the game or to use the computer. Whatever remains to be learned is immediately put into relief.

Here is a suggestion for a first or early experiment. Use two to five acres, though a smaller area would also be useful. Make it a weed control, rather than a fertility, experiment, since no alternative fertility will yet be present. Use soybeans. This will demonstrate how soybean plants tolerate aggressive weed control procedures and whether yields are adversely affected by reduced population. If possible, include both sloping and level ground or both terraced and unterraced areas in order to see how weed control differs in these contrasting situations. Lay out the plot so that weed control implements can be used properly. For example, a long, narrow plot will allow a rotary hoe to work effectively. If possible, plant half of the plot 7 to 14 days later than the first half in order to observe the differing weed control situations. For example, keep track of the times of emergence for the beans and the weeds. If early weed control fails, retain the plot. Make it an opportunity to learn about other aspects of nonchemical weed control. For example, do selective hand roguing, such as leaving the pigweeds on half of the plot, to see how yields are affected. Or replant late to see how this works out.

An experiment like this will provide quick information and valuable experience in such things as terrain, conservation structures, planting dates, machinery settings, appropriate procedures, and timing with respect to both weed control and other objectives.

There are two additional management strategies which are important in the preparation for and early stages of the conversion process. One is the careful assessment of soil fertility and initiation of measures to improve it. An extensive program of soil testing should be instituted and maintained. Testing provides information about soil resources such as organic matter, acidity, and fertility.

Early soil tests may reveal that soil pH is out of balance, a condition resulting from the years of steady application of chemical fertilizers. If the tests so indicate, do not hesitate to apply lime. For one thing, the crop rotation will be incorporating more and more legumes, which are net consumers of calcium.[7] Furthermore, if during the gradual conversion the farmer continues using nitrogen fertilizers on acidic soils without applying lime, the acidity may in fact increase. According to Miller and Donahue, "Crops respond to liming because it improves microbial activity, it corrects calcium, magnesium, and molybdenum deficiencies, it lessens aluminum, manganese, and iron toxicities, and it makes phosphorous more available and potassium more efficient in plant nutrition. . . . The continued use of ammonium nitrogen fertilizer on acidic soils without application of lime may *decrease* soil productivity by making the soil *more acidic*" (1990, 226–27).

Another important early management strategy is to write a detailed plan for the conversion. This will be very beneficial not only to the farmer, but also to others associated with the farm, such as a landowner or a lender. The plan should include a schedule of changes for every aspect of the operation. The farmer can then confront variances from the timetable with a clear sense of what to do next. For example, suppose that a field is scheduled for complete conversion in a specific year. However, the farmer has realized that a cocklebur (*Xanthium strumarium* L.) problem must first be brought under control. The timetable and overall plan can provide the context for reviewing the situation. As a result, a farmer might then decide to alter the plan's placement and scheduling in order to put the problem field into alfalfa immediately.

These, then, are some of the key decisions and plans which must be made and management strategies which must be implemented as the process of converting to chemical-free farming begins. I can't stress enough the importance of careful background preparation to attaining the goal. Good preparation and a gradual approach are crucial. Of course, there will be additional considerations for persons other than an owner-operator, such as a tenant farmer, a non-operating landowner, or a lender. I will discuss some of these later in this chapter. However, now I wish to turn to the central feature of the conversion: the reduction and elimination of insecticides, herbicides, and synthetic fertilizers.

Reducing and Eliminating Chemical Use

The key to controlling insect problems without using insecticides is careful selection, placement, and rotation of diversified crops. Several factors are relevant to insect management. With the exception of alfalfa, the same crop will seldom be grown for two years in a row on the same field. As a result, the reproductive cycles of insects will be interrupted. For example, not repeating soybeans on the same field helps to control cyst nematodes.[8] Another controlling effect stems from the fact that new crops will be part of the rotation. For example, a new crop such as milo in the feed grain part of the rotation will disrupt the cutworm, grasshopper, and corn borer cycles that are associated with corn. A third factor is the smaller field sizes which result from incorporating more kinds of crops in the rotation. Smaller fields make it possible to contain insect infestations more easily and effectively. And finally, the greater variety of crops and complexity of their placement provide options for bordering one crop with another for purposes of insect control. For example, a chinch bug problem in wheat will be better contained by placing a crop like soybeans beside it rather than a natural chinch bug host such as forage sorghum.

In general, eliminating insecticides will be the easiest part of the conversion. In the 12 years I have been farming without them, I have had almost no insect problems. I once lost half an acre of corn to cutworms, but promptly replanted it. Occasionally the next cutting of alfalfa is slightly delayed by weevil, and there are still mild greenbug infestations. When I began full-time farming in 1975, one farm had a chronic greenbug problem. But once a diversified crop rotation was in place—and once we quit spraying—ladybugs repopulated the fields and the greenbug population became manageable. It also helps to select plant varieties with good insect resistance.

There is a final important aspect of managing insects without insecticides. Farming with livestock provides an additional option for coping with severe insect problems. If I were to lose a field, I would replant it to a forage crop appropriate to managing the insect problem. Possible choices might be turnips, triticale, rye, and forage sorghum. This could potentially provide several tons of forage per acre for fall and winter feeding. Such an option reduces the financial consequence of insect devastation.

A farmer who has been following the system of conversion which I describe, one in which a diversified crop rotation is in place, will be well positioned for the final severing of herbicide use. The pressures to use herbicides will be greatly reduced. A field infested with shattercane may be placed in wheat or alfalfa rather than corn or soybeans. Problems with pennycress (*Thalaspi arvense* L.) can be effectively addressed by substituting fall-maturing row crops for sod-base crops. A field with hemp dogbane or morning glory infestation may be planted to oats rather than to a row crop in order to permit summer moldboard plowing, which is extremely effective against these weeds in dry weather. These are just a few of many such possibilities.

Because of the importance and complexity of managing weeds without herbicides, I discuss the system for doing so in great detail in chapter 3. Here I will simply suggest some considerations for choosing the first full field without herbicides, as distinct from the first experiment. Don't select a field with a serious weed problem. For example, a field with a history of velvetleaf (*Abtilon theophrasti* Medic.) infestation will present great difficulties. A good choice would be a field with an allelopathy advantage, for example, one on which the prior year's crop was alfalfa, rye, oats, milo, or turnips. I also find it easier to control weeds in the second year of successive years of soybeans, though I don't yet understand why. Try to avoid choosing a field with a large amount of residue. Residue management can be postponed. And don't launch the first herbicide-free field until there will be time to pay attention to it, for unlike treated fields, it will hardly be on automatic pilot.

Achieving the goal of reducing or eliminating synthetic fertilizers does not require different steps from those already implemented in earlier stages of the conversion process. At the final stages, the crops will be benefiting from rotation, meticulous liming, and animal and green manure. Further reductions in synthetic fertilizer use will flow from specific decisions about feed grains, legumes, yield goals, and livestock. Reduced acreage in feed grains such as corn implies a reduced overall nitrogen requirement for the farm. Legumes in the rotation contribute to nitrogen supplies, in addition to providing other benefits to fertility. Yield goals sometimes change when farmers notice that top yield does not always translate into highest net profit. Finally, greater numbers of livestock provide more manure and make more legumes viable.

The process of conversion which I am describing is a deliberate one, requiring five or more years. To my mind that is one of its virtues. However, some farmers, while sensible enough not to convert all at once, may be in a greater hurry. For them I offer these observations, suggestions, and cautions. First, don't increase the size of the initial experimental plot; as much can be learned on 5 acres as on 40. Second, it is quite likely that herbicide carryover will be a limiting factor in rotation or weed control plans for the first year or two. The fact that this must be worked through is a strong argument for converting slowly. Third, expand the percentage of acres in alfalfa, oats, and wheat first. These crops tend to make other reduction steps easier, and they work out well financially on farms where there are sufficient numbers of livestock. Fourth, one way to temper impatience with the slowness of conversion is to devote time early in it to labor-intensive tasks such as tree planting or fence building. And finally, consider participating in the long-term federal Conservation Reserve

Program. The acres of grass will not require the quantities of pesticides used in conventional farming.

Institutions

In the conversion to chemical-free farming, the on-the-farm process itself presents sufficient problems to challenge and occasionally discourage even the most dedicated farmer. However, on top of these there are nonagronomic obstacles which, in many cases, may be even greater impediments. Among these are certain features of the federal farm program and, for some farmers, the need to convince and work with lenders, farm managers, and farm owners.

FEDERAL FARM PROGRAMS

The federal farm program that emerged from the 1985 and 1990 Food Security Acts created an extraordinary obstacle to conversion to pesticide-free farming. It is the notorious "use or lose" feature of the crop subsidy program. When a farmer plants less than his or her permitted acreage, a formula comes into play that can reduce permitted acreage—the basis for program benefits—in subsequent years. As stipulated in the 1985 legislation, the acreage base equals one-fifth of the total acreage planted over the previous five years. Thus, for example, a year of rotation to a nonprogram crop such as clover could result in the loss of 20% of a farmer's base. In 1990 the program was amended to permit some substitution of nonprogram crops on base acres without loss of base. For example, the Integrated Farm Management Program permits some soil-saving crop rotations, and the Water Quality Incentives Program provides incentives to protect water quality in targeted areas. But the basic problem still exists.

Since pesticide-free farming usually involves less and less reliance on the federal farm program, there are ways that the converting farmer can minimize the impact of the "use or lose" feature. On a typical midwestern farm in a soybean and corn rotation, the feed-grain base is usually 50 to 60% of the total acres. However, with the crop diversity and rotation required for pesticide-free farming, such a large feed-grain base cannot be preserved indefinitely. The key is to minimize the reduction for as long as possible. This can be accomplished by finding the necessary acreage for rotation and diversification in nonprogram crops. For example, space for alfalfa can be taken from soybeans, which are not a program crop. Only after the nonprogram acres have been pared down as much as possible will the feed-grain base need to be reduced, and this can be spread out over many years. A farmer should allow time to become confident with the new crops, rotations, and methods before giving up much of the feed-grain base.

A positive aspect of the federal farm program is that it provides attractive opportunities in two of its set-aside programs. The yearly set-aside of the Acreage Conservation Reserve can be used to begin crop diversification and rotation, to isolate and experiment with special weed problems, and to establish legumes and grasses. The program can also provide forage opportunities. The Conservation Reserve Program, a long-term set-aside, can be used to establish grasses and valuable trees, and it is a way to help stabilize income. Because of strict weed control requirements for lands in these programs, a farmer wishing to meet these provisions other than with pesticides should make these arrangements before signing the contract.

LENDERS, FARM MANAGERS, AND FARM OWNERS

In addition to needing to accommodate to the federal farm program, the farmer converting to an alternative system may well have to persuade lenders, the farm manager, or the farm owner of the wisdom and viability of chemical-free farming. While they may be sensitive to its environmental benefits, it is safe to assume that economic considerations will be decisive.

A subject to emphasize with a lender is the financial advantage of livestock. Livestock projects will almost surely be introduced, expanded, or merely more carefully organized as the farmer carries out the conversion process. At minimum, this offers the promise of greater financial stability.

Issues relevant to lenders, managers, and farm owners are the prospects for reduced costs and enhanced financial stability through diversity. Cost reduction arises from gradually trading management for cash inputs and from using methods that economically enhance soil resources, such as crop rotation and use of legumes and grasses.

Lenders, farm managers, and farm owners are entitled to considerably more than the general discussion suggested above. Consider what a conversion-minded farmer is proposing. The proposed system is fraught with uncertainties, and it gradually but permanently reduces government financial support. Even if the changes can be made plausible, lenders, managers, and owners must decide whether the individual farmer who would be making the change is capable of doing so.

To make the case specifically, the farmer should share with these people the written plan he or she has prepared describing in detail the conversion process and the timetable for it. The timetable should provide for periodic reviews of the plan and its implementation with all interested parties. For the benefit of a farm owner, the plan should also emphasize the beneficial changes to the farm, such as improved soil resources. Another advantage that can be presented is that a set of buildings designed for earlier, more diversified farming will likely be quite suitable for the system being proposed. It also might be helpful to present evidence of the effort put into learning about sustainable agriculture, such as listing background readings, seminars and meetings attended, and tours taken. Some of this material could be shared.

However, it must be acknowledged that tenant farming presents serious obstacles for conversion to pesticide-free and sustainable farming. One issue is establishing terms that protect the tenant's situation. For one thing, conversion requires work, sacrifice, and the acquisition of skills, all of which pay off only toward the end of the process. For example, a typical long-term soil-building program will include detoxifying, adjusting acidity, applying manure, and instituting a crop rotation program. The positive results, such as increased organic material and improved tilth and water absorption, will be years away. Furthermore, in the conversion process, capital inputs, part of which the farm owner supplies, are replaced by labor and management, most or all of which the tenant provides. For example, chemical weed control may be replaced by a strategy that includes careful rotation, carefully chosen planting dates, rotary hoeing, harrowing, and shovel cultivation of row crops.

A minimally acceptable lease should address these matters directly. It should be long-term, assuming that the tenant performs as agreed. It should specify the landowner's responsibility for infrastructure investments, such as fences and grain storage. And it should provide a means of compensating the tenant for adopting methods that partially replace costs to the landowner, such as adjusted crop share or rent or cash reimbursement. In general, if the landowner agrees only reluctantly to trying sustainable techniques or is not willing to tolerate experiments and mistakes, conversion is not likely to succeed.

We have been considering conversion largely from the perspective of the farming operator. I turn now to the situation of the absentee farm owner. Imagine someone who owns a farm operated by someone else, who has an interest in land stewardship, and who has a suspicion that the farm is not handled as well as it could be. What can be done? Over half of the farm land in the United States is farmed by nonowners (Lewis 1978). Often farm owners in such situations who care about their farms see themselves as being without recourse, limited by the priorities and dedication of their tenant. I hope to show that some alternatives are possible.

An uncooperative tenant can frustrate every move towards stewardship. A stewardship-oriented farm owner in such an unfortunate situation must make a choice. One of the inspirations for this subject, however, is the conviction that many tenants who are otherwise silent about general priorities for a farm would respond positively to stewardship-oriented initiatives by the farm owner. Implementation of stewardship ideas presupposes informed agreement between the farm owner and tenant.

The first subject is soil conservation. Consider permanent land structures. Mainly, the structures involved are water diversion terraces and a network of grassed waterways. The federal Soil Conservation Service provides technical assistance at no charge. This process can begin with a meeting of the Soil Conservation Service technician, tenant, farm owner (and farm manager, if involved) at the farm to identify needs and establish priorities.

Major conservation work should be spread over several years. The tenant must not be overwhelmed with the management of new structures. New terrace channels require soil amendments and special tillage, waterways must be seeded, and new farming techniques may have to be mastered. When waterway construction is spread over several years, the risk is minimized that a whole system of new waterways will be assaulted by severe weather before seeding is established. Also, Soil Conservation Service cost-share funds for these structures are, at best, available each year on only a limited, competitive basis. Possible tax advantages may also be considered.

Establishing priorities does not necessarily consign parts of the farm to continued erosion. Eroding fields that cannot be scheduled for immediate work may be placed in government set-aside programs or planted to crops which minimize soil erosion, such as small grains. Eroding fields which cannot be improved for several years may be seeded to alfalfa, which will largely arrest the erosion problem. The landowner will be required to make several decisions. How large will he or she permit the terraces and waterways to be? Larger structures, which are more expensive, promote longevity, function, and compatibility with larger equipment. Will the terraces be placed parallel to each other? This procedure, usually more expensive, eliminates point rows (crop rows which fill out spaces other than squares and rectangles), which can improve yield and make terraces which are less frustrating.

Basic farm assessments often yield dubious proposals. I suggest that, if a farm owner is advised to remove trees, drain wetlands that are not already protected by law, straighten creeks, or have grasslands plowed up, a second opinion be sought. A local organic agriculture organization or Sierra Club, for example, should be able to refer the landowner to a qualified person who can provide an alternative opinion about the economic viability and environmental impact of such actions.

An additional advantage of a farm care plan which begins with soil conservation is that compliance with federal conservation regulations is assured. The regulations stipulate that any farm participating in government programs must have had a soil conservation plan by 1990 and a viable conservation system established by 1995.

Key to a successful owner-tenant partnership for stewardship is the demonstrated interest of the farm owner. This means sustained communication, including regular visits to the farm. The tenant can hardly be expected to take more interest in the farm and stewardship of it than does the farm owner.

One way to foster a sense of partnership as well as to advance the tenant's skills in stewardship-oriented agriculture would be for the owner to strongly encourage and support on-farm experimentation. In addition, the owner should consider subsidizing the tenant's continuing education in alternative agriculture, such as through relevant conferences and seminars and through subscriptions to appropriate magazines and journals.

The farm owner might use incentives to motivate the tenant to rotate and diversify crops. For example, suppose the owner wants to incorporate wheat, oats, and alfalfa into the farm's operation. The owner can offer to assume a greater portion of the costs, to increase the tenant's share, or to reduce cash rent on acres planted to these crops.

Granted, it is difficult to create an interest in livestock. Suppose, however, that the owner actively cooperates in putting livestock on the farm. This might begin by the owner agreeing to pay for fencing and basic handling and watering facilities. The value of the tenant's labor could be amortized in the lease so that it would be partially repaid if he or she left the farm. Also the landowner could permit the tenant to have all of the wheat and oat straw, provided that it is used on the farm in livestock production, thus promoting nutrient recycling.

The main role of the landowner in reducing insecticide use is to be supportive. As soon as crops have been diversified and placed in rotation, the tenant can be encouraged to experiment with all forms of insecticide reduction. The landowner should make it clear that he or she will share the consequences of experimentation without complaint.

Herbicide reduction is a much more ambitious step and requires a tenant with dedication and perseverance. A landowner who is lucky enough to have such a tenant can assist in several ways. Education about alternative methods of weed control should be encouraged. Most important, the landowner must be willing to share costs. The tenant should be compensated for the greater work and management required, either in the general terms of the lease or on a per-acre basis. Such costs are in lieu of the costs of chemicals.

In general, cash rent will not be conducive to establishing a stewardship-oriented farming operation. Presumably a landowner and tenant with shared interests in the stewardship of the farm will think of themselves as partners in pursuit of that goal. Partnership implies the sharing of gains and losses; cash rent does not. If, for whatever reasons, the farm owner must have cash rent, he or she will have to make special efforts to demonstrate interest in a sustainable operation of the farm.

Attainment of stewardship goals can also be complicated by the use of a farm management service to oversee the farm. However, if the owner finds that this is necessary, it is vitally important to contract with a farm manager who is not just grudgingly receptive to the stewardship goals but is enthusiastic about them and is knowledgeable and experienced in the ways of pursing these goals.

Beginning Farmers

If careful planning and implementation are important to the established farmer converting to chemical-free farming, they are even more so for the beginning farmer aspiring to farm without pesticides or synthetic fertilizers. The new farmer must not only deal with the components which I have just described, but must also make a number of other important planning decisions, ones which may significantly affect the viability of the venture. In making choices about the variables I discuss below, the guiding principle should be to postpone for as long as possible commitments—especially financial decisions—which limit flexibility.

One such decision will be the method of planting row crops. Both ridge till and surface planting are popular with farmers using alternative systems. Each has distinct advantages and disadvantages. I suggest that part of the information-gathering process should be to visit well-managed farms which use each method. It is important to note that selecting ridge till creates pressure not to construct terraces because of the incompatibility of ridges and the point rows associated with terraces. This is especially important for farms with highly erodible land. If the mandatory soil conservation plans for such land do not include terraces, tillage methods and permissible crops are much more restricted.

Many kinds of livestock will work on a pesticide-free farm. Beef cow-calf operations offer a number of advantages. For one thing, the investment in buildings and equipment is generally lower than for other livestock, and much of what is needed—barns, silos, fences—may already be present. Furthermore, the operation can most likely be easily be handled with family labor. Another advantage is that a beef cow herd is less speculative than some other kinds of livestock raising. Market prices are less likely to fluctuate greatly, and there will be less opportunity and temptation to try to outguess the market by jumping in and out of production (Neumann 1977, 212–13).

Because farm machinery is so expensive and because a farmer's capital is so scarce, decisions about machinery purchases are very consequential for both established and beginning farmers. Buying just one costly piece of equipment which is not effective in pesticide-free farming can make the commitment to it much more difficult. And it is important to keep in mind that often the implements or models that conventional farmers have made popular are not the correct ones for pesticide-free farming.

One strategy for avoiding expensive machinery purchases is to hire custom operators for some jobs. In addition to preserving scarce capital, custom hire has several other advantages. It is a good way to address a labor shortage, especially in peak workload periods, and it may enable a farmer to retain a job off the farm. Furthermore, it provides the time and opportunity to see what works best. For example, a farmer might want to decide whether to own a baler that makes round bales or one that makes square bales. Experience with both will help to decide. Additionally, the cost per project for custom work is seldom more than the cost associated with ownership. One potential source of savings is that it postpones the need to provide inside storage space for the implement.

Do not buy one of the conservtillage implements that have become so popular for fall tillage, that is, machines that slice residue, till, and leave a high percentage of residue on the surface. In the system of farming which I am describing, fall tillage will be virtually eliminated as a result of the cropping diversity, the soil conservation goals, and the use of the fields by livestock. And in the spring there is scant opportunity for single-pass tillage because of late planting dates.

Another implement to avoid or delay purchasing is an expensive grain drill such as those which are popular for drilling soybeans on conventional farms. Generally, farmers who don't use pesticides prefer to plant soybeans in rows. In addition, small-grain crops are usually planted in less busy times of the year. Therefore, the quality of the drill is less important. And because drilling is not a difficult procedure, it is well suited to assistance by hired help.

Although larger and larger trucks are popular on conventional farms, there are numerous reasons to postpone or avoid this especially expensive investment. On a diversified farm much of the crops will be cycled through livestock rather than marketed directly. For hauling livestock, trailers are probably preferable. And when grain is sold, it is quite likely that local grain elevators will offer competitive prices to pick it up at the farm. Furthermore, farmers who develop special organic markets or who sell premium hay often sell commodities several hundred miles (and several states) away, which entails a major trucking commitment.

In addition to equipment and tillage decisions, the beginning farmer will want to consider what facilities are needed. In recent years it has been popular to "clean up" farmsteads by taking out fences and groves and tearing down older buildings such as granaries, barns, and corn cribs. One reason for this trend is that conventional farming appears to have rendered these facilities obsolete. However, on a pesticide-free farm, the greater complexity of the operations makes necessary a more varied set of facilities. Therefore, the farmer beginning in or converting to an alternative system should consider how the existing infrastructure can be adapted to meet present needs. Dilapidated fences often have enough good wire and posts to justify repairing them. Old granaries provide good storage for feed, oats, and seed. Old horse barns, with their lofts, internal granaries, and stalls, are ideal for calving. Old corn cribs often have very convenient overhead bins, their side cribs can be easily modified for hay storage, and the center alleys have a variety of uses. Even improbable-looking stands of trees can make valuable livestock shelter, especially as more satisfactory windbreaks are being established. Hence, overly ambitious cleanup can deny farmers some of the facilities needed for pesticide-free farming.

The grain storage needs of pesticide-free farms differ fundamentally from those of conventional farms. Generally, more and smaller storage units are preferable to a few large ones, to accommodate diversified cropping.

Conclusion

The challenge of converting to pesticide-free farming turns out to be rather complex. The federal farm program sets forth profound obstacles. The agronomics of conversion are also a challenge. All players—the farmer, lender, manager, and landowner—must work together as a team. The key is to recognize these factors at the beginning and prepare accordingly. This chapter has sought to contribute to that preparation.

3

• • •

Weed Management

Central components of nonchemical weed management are crop rotation, tillage, aspects of planting, and mechanical weed control.

Crop Rotation

The structure of crop rotation is not rigid. There are several successful formats. The guiding principle is to alternate crops in ways that help to achieve specific objectives. This will mean much of the following:

- Alternate sod-based crops, such as wheat, oats, and rye, with row crops, such as corn, milo, and soybeans in order to combat both weeds and insects. Crops tend to promote weeds with similar life cycles. For example, wheat and oats help weeds such as pennycress and downy brome (*Bromus tectorum* L.) to thrive. Corn and soybeans complement weeds such as velvetleaf, smooth pigweed (*A. Hybridis* L.), cocklebur, and shattercane (*Sorghum bicolor* [L.] Moench subsp. *drummondii* (Steud.) de Wet). Therefore, planting a field infested with weeds associated with sod-base crops to row crops permits tillage and cultivation at times when the weeds are vulnerable. And planting a sod-base crop in a field infested with weeds associated with row crops will allow tillage and/or harvesting at times when the weeds can be set back.
- Alternate weed-suppressing crops, such as alfalfa, rye, oats, milo, and turnips, with crops which lack this characteristic, such as corn.
- Alternate crops which are hosts to specific insects with crops which do not serve as hosts. For example, in my rotation, sorghum, which is susceptible to greenbugs and chinch bugs, is commonly followed by soybeans, which are not.
- Alternate soil-enhancing crops, such as alfalfa and grasses, with soil-

depleting crops, such as corn. For example, corn negatively affects soil texture and consumes nitrogen, whereas alfalfa improves tilth and fixes nitrogen. On the other hand, crops which absorb nitrogen reduce chances that surplus nitrogen will leach into ground water.[1]

Fortunately, several crop pairings satisfy most, if not all, of these considerations. Two such rotations are grass with soybeans and alfalfa with corn. A rotation should commence with a soil-enhancing crop.

Tillage

In general, spring tillage which contributes to pesticide-free farming and which meets soil conservation requirements has the following characteristics: It is diversified, delayed, and minimized. Furthermore, it preserves maximum residue from the previous year's crop, and it compacts the soil as little as possible.[2] To illustrate how these goals can come together, I will describe my spring tillage routine.

For early- and mid-spring tilling, I use two implements: a moldboard plow (on a very limited basis) and a disk. I use the plow to destroy old alfalfa fields, to reinforce terraces, and to address special weed problems, such as common morning glory (*I. purpurea* [L.] Roth) or bindweed. I use a disk to work the problem weed areas that have been plowed, to condition newly plowed alfalfa fields, to incorporate manure, and sometimes, when residue is heavy, to till for oats. I often pull a rolling packer behind the disk when tilling newly plowed alfalfa fields to break up clods more effectively and to conserve moisture. Because of the broadly diversified cropping, these operations can usually be carried out in a leisurely way, thus avoiding the temptation to work wet soil and also allowing plenty of time for other important activities, such as monitoring calving.

I usually begin tilling for row crops sometime between late April and mid May. I prefer not only to have avoided fall tillage but to have delayed spring tillage for as long as possible. It begins when weeds have grown to the point that, if they were any bigger, a single tillage would not reliably destroy them or, in the case of pennycress, it would go to seed. Late spring tillage has a number of important benefits. One is that a crop of weeds can be destroyed with the very first tillage. Second, delaying tillage minimizes the succeeding weed control operations, thus reducing not only costs but also the deleterious effects of tillage on soil.[3] Third, late tillage allows for further weathering of surface residue, making it easier to manage.[4] And finally, the schedule extends grazing and calving options for cattle producers. For example, my cattle graze through the winter and sometimes into April. And undisturbed stalk fields have also been useful locations for calving, especially in wet periods.

In my late-spring tillage I avoid disking and plowing as much as possible, preferring instead to use a field cultivator. Sometimes it is used for the first spring tillage of stalk fields. The strategy, which can be thought of as ultra-minimum tillage, has several conservation benefits. Because of the way the field cultivator works, shredding, cutting, and breaking of stalks is minimized. At the same time, field cultivating leaves 50–80% of the residue on the surface, more than any other tilling method (Dickey, Jasa, and Shelton 1986, table 1; U.S. Soil Conservation Service 1992, 18–19). Another conservation consideration is that a field cultivator eliminates the erosion-promoting ridges associated with most tandem and some offset disks.

Field cultivating also has several practical advantages over disking. A field cultivator has a lower horsepower requirement per foot of implement, and it can be used at higher speed. In my experience, a field cultivator may be pulled two to three miles per hour faster than a disk.[5] Furthermore, a field cultivator does a better job of leveling the ground, and it is easier to maintain.

Stalk residue should be worked with a field cultivator in the driest part of the day. When the machine does plug up, it should be unclogged gradually to scatter the stalks over a larger area. I do this by slowly raising the field cultivator as I drive ahead or turn around so that the residue is not dumped in one big pile. Plugging of the harrow sections at the rear of the machine can be minimized by adjusting them so as to reduce the downward pressure.

In my own situation, direct field cultivating of corn stalks has consistently worked well. I have had very little plugging, even though all my tillage must be done on the contour between terraces in the same direction as the rows of the previous year's crop. Likewise, field cultivating has worked well on soybean stubble. In situations where seeding can be done after one pass, much more residue remains on the surface than would be the case with disking. However, I have had some problems with direct field cultivation of milo stalks. As a result, I usually field cultivate those locations where I can drive at an angle to the rows, and I lightly disk the remainder. Then subsequent tillage of the whole field can be with the field cultivator. This plugging situation with milo might be reduced by more intensive grazing or with an improved field cultivator. Mine is a 1975 Krause, which has a frame length of only 56 inches, allowing little residue clearance from the rear to the front row of shanks. More recent designs are much improved. For example, frame length of the Wilrich 10 FCW is 90 inches, the Case IH 4600–4800 series frame length is 130 inches, and the John Deere 9605 is 89 to 135 inches.

The final tillage before planting should be carefully timed and executed. It should occur at a stage of weed growth when one pass will be effective. It should be thorough; weeds not uprooted will survive later hoeing, harrowing, and possibly even cultivation. The shovels on a field cultivator should be sufficiently wide to destroy difficult weeds, such as cocklebur. The thoroughness of the tillage on uneven ground, such as in rounded terrace channels, is improved by using a flex-wing field cultivator, one with a narrow center section and wings. However, tilling at a uniform depth with a flex-wing cultivator requires careful adjustment of the linkage between the center section and the wings.

The final preplanting tilling should not be done when the soil is wet. In addition to contributing to ground compaction, tilling in such conditions can stimulate germination of weeds like velvetleaf and pigweed, which, given a head start, will be difficult to uproot later.

As much as possible, the final tillage should avoid leaving tracks. For one thing, weeds are more likely to germinate there. Also, in certain locations, tracks contribute to water erosion. And tracks can interfere with the proper functioning of planters, especially ones with worn furrow-opening parts.

In the final tillage, special attention should be given to preserving the structural integrity of waterways. One problem with the newer, longer-framed field cultivators is the difficulty of getting uniform tillage at the place where a round of tillage begins, such as at the edge of a waterway or a field. Thus, there is the strong temptation to obtain uniform depth by tilling parallel to the waterway. However, this has the potential of inducing erosion beside the waterway, effectively destroying its function. If it is necessary to till parallel to a waterway, do so at the beginning of the operation so that the remaining rounds done perpendicular to the waterway will partially erase the patterns that run parallel to it.

Planting Strategies

Much of what is required for successful nonchemical weed management in row crops takes place at and even before planting. One unifying principle is creating and sustaining a competitive edge for the crop.

Table 2. Yield and days from planting to maturity of specific soybean varieties at different planting dates at Mead, Nebr., 1976–1978

	Yield (Bu/A)			Days to maturity		
	Wells	Woodworth	Cutler 71	Wells	Woodworth	Cutler 71
Planting date						
Early–mid May	36	40	36	124	136	141
Mid–late May	39	35	32	117	118	130
Early–mid June	37	32	27	105	111	118
Mid–late June	31	28	23	99	104	109
Early–mid July	19	18	14	91	93	93

Source: Elmore and Flowerday 1984, 2. Based on results reported in Essa 1979 and Johnson 1979.

TIMING OF PLANTING

The timing of planting is an important factor, especially for weed control in row crops. In general, farmers who do not use herbicides tend to plant later than those who do. There are at least three advantages. One is that warmer soil temperatures speed the emergence of the crop. For example, I have had early-planted soybeans take as much as 14 days to emerge whereas late-planted ones have come up in as little as 3 days. Faster emergence provides an important competitive edge over weeds. Second, later planting allows for later tillage, which is especially effective in controlling early-germinating weeds.[6] Third, in some areas, such as in eastern Nebraska, rain is generally less frequent and of shorter duration later in the planting season.

In my operation in eastern Nebraska, the general schedule for planting row crops is

- corn after May 15
- sorghum close to June 1
- soybeans after May 20 and into June

Support for such relatively late planting of soybeans can be found in a variety of authoritative sources. For example, in their standard text on crop production, Martin, Leonard, and Stamp (1976) say that "soybeans are best planted in May or early June in most states. . . . Earlier planting hastens flowering and reduces yields, and in the north, the soil is usually too cold to plant soybeans before late May or early June" (702). And in their summary of research conducted in eastern Nebraska, Elmore and Flowerday (1984) conclude that "soybeans have a unique ability to yield well when planted over an extended time period." In Table 2, I reproduce data which Elmore and Flowerday report. Notice that, with the exception of the Cutler 71 variety, yields did not decline precipitously until after mid-June planting dates.

The results with Cutler 71, a longer season variety, illustrate the need to select somewhat shorter-season varieties to accommodate later planting dates. My experience is that recent short- and medium-season varieties have excellent yield potential.[7]

STAGGERED PLANTING

A particularly useful planting strategy to incorporate into pesticide-free cropping is to space out the dates of planting different fields. This can be characterized as planting as slowly, rather than as fast, as possible. There are several reasons to stagger plantings, especially soybeans, which have much more flexibility for later planting. The first reason is that it maximizes the chances that the farmer will be available to do mechanical weed control exactly when it is needed and is most effective. Consider the opposite situation, the common one in which planting is completed as fast as possible. Even if this involves only a few hundred acres, the farmer quickly finds it impossible to do all the nonchemical weed control procedures at the time they are needed. That would be an especially foolish and avoidable way to compromise weed control. In my operation, I try to spread out the planting of approximately 175 acres of soybeans over about a three-week period, with the last planting saved for fields with greater weed problems. At the minimum, do not plant more in one day than can be properly rotary hoed in one day.

A second reason to stagger planting dates is that it spreads the weather risks. For example, an extended wet period compresses work schedules, creating a need to do more things at a particular time than was originally planned. However, by having a variety of field situations, a rainy spell is less likely to be such a serious threat to effective weed control. And having fields staggered from earlier to later planting reduces the amount of exposure to an early frost.

A third advantage of staggered planting is that it spreads out the work load at harvest time. This is especially important for soybeans because they should be harvested when they reach the optimum moisture at which field loss and bean damage are minimized. Even when several varieties have been used, sometimes more bean fields reach and pass the optimum than can be immediately harvested. Staggered planting helps to cope with the problem.

To illustrate the effect of planting dates on weed control, let's consider two fields of soybeans, one planted relatively early and one relatively late. Suppose a field of soybeans is planted May 20 in eastern Nebraska. Based on my experience, they will not emerge until 10 or 11 days later. That gives certain varieties of weeds ample time to germinate and to emerge ahead of the crop. For someone farming without herbicides, such a situation means that a preemergence hoeing will have to be done, say on May 25. The beans will not come up until May 30 and cannot be hoed again until 2 or 3 days later. Thus 7 or 8 days have elapsed since the first hoeing, ample time for rain and a second surge of weed growth before the crop is even established.

Now suppose that a field is planted on June 7. According to my experience, the crop will come up in 3 or 4 days, partly depending on variety. (Fast emergence is an important consideration in selecting soybean varieties.) Thus, the crop will usually emerge before the weeds do, even before fast-paced velvetleaf. In this situation simply wait to begin hoeing until weeds are barely visible. Several factors affect how long the operator waits:

- how careful the preplanting tillage was
- how effectively the moisture management was at planting
- how long rain is delayed after planting
- whether an allelopathic situation is present
- whether weeds are simply delayed, a situation which occurs unpredictably for reasons which I do not understand

Waiting has several important benefits. The longer hoeing or harrowing is delayed after emergence, the less seedling loss there will be because the crop is better established. Second, the better established the crop is, the more aggressive the hoeing can be, resulting in greater weed control. Third, as the cultivating stage draws near, the chances steadily increase that one hoeing will be enough. And finally, fewer hoeings reduces ground compaction, a condition which, among other things, can interfere with effective cultivation.

Thus, simply planting soybeans later has the distinct advantages of maximizing weed control, preserving stand, and reducing work. Yet I have also stressed that staggering the plantings is important. Obviously, these two strategies cannot work in perfect harmony, since in order to gain the benefits of staggering, the planting must start earlier. One effective response is to do the earlier planting in less weedy fields or those benefiting from allelopathy and to plant later those fields which are potentially weedier. In 1990, one of my early-planted fields was rained on four times between planting and the first rotary hoeing. However, I was able to retain weed control because of the allelopathy provided by the previous year's milo crop.

Factors which affect emergence time, assuming sufficient moisture, include

- soil temperature
- planting depth
- seed variety
- the quality of seed-to-soil contact

MANAGING SOIL MOISTURE

A crucial aspect of nonchemical weed control is managing soil moisture at the time of planting. The governing principle is to plant in a seedbed which is as dry as is consistent with uniform germination. I will now explain what this means and why it is important by examining a range of moisture possibilities and specifying their consequences for germination and weed control.

• Much too dry. If the soil is much too dry, virtually none of the crop will germinate without rain. The problem is that the rain which germinates the crop will also germinate weeds. Everything comes up at the same time, giving the crop no advantage.

• Too dry. If the soil is somewhat more moist, but still too dry, part of the crop will germinate and sprout, but part of it will not until it rains. The problem is that all subsequent weed control—hoeing, harrowing, cultivating—presupposes a size difference between the crops and weeds. In this case, however, only some of the crop will be larger than the weeds; the rest will be the same size.

• Too wet. At the other extreme, planting in soil that is too wet can also create problems. Surface weeds, especially grasses, velvetleaf, and pigweed, germinate quickly, often sprouting even before late-planted crops. Indicators of conditions too wet for planting are visible moisture on surface soil after the planter has passed through it and very distinct tracks left by the furrow-closing wheels.

• Just right. Conditions are optimum when there is adequate moisture below the surface to germinate the crop, but the surface is dry. This situation creates a significant competitive advantage for the crop by permitting it to germinate and sprout immediately, while delaying germination of weeds until the next rain. And the longer it is until it rains, the greater is the competitive edge for the crop.

Of course farmers cannot control the weather, but there are ways to greatly increase the chances of having advantageous moisture conditions at planting. One is to use a recent-model planter. The latest generation of planters, such as the John Deere Maxemerge and the Case IH Early Riser, have features that can assist in moisture management. The furrow-closing devices have become more sophisticated. Whereas older planters have shovel-like disks which scoop and heave covering soil, the newer equipment gently nudges or squeezes soil over the seed. This procedure is more successful at leaving available moist soil around the seed and the drier soil on the surface.

Another component of managing for optimum moisture is to practice patience, to learn to wait until conditions improve rather than rushing out and planting when it is merely physically possible. Suppose, for example, that rainfall has been plentiful. The first step is to let the fields dry out before the final preplanting tillage. Working wet soil contributes to compaction, results in poor contact between seed and soil, and causes weeds to germinate. Next, allow still more time to elapse between the final tillage and planting to fine-tune seedbed moisture. In wet conditions, I have waited from 24 to 36 hours between final tillage and planting. For farmers converting from conventional farming this approach will be an adjustment of practice, since herbicide usage allows working wet fields without paying a penalty in weed control.

On the other hand, suppose that conditions are very dry. In general, it is best to wait for rain and then let the fields dry out properly before planting. However, don't allow weeds to grow excessively. This will sap moisture and create the risk of having to till twice before planting, further depleting soil moisture. In dry conditions, till in a manner that minimizes double tillage. Till field ends last. Finish areas other than squares and rectangles by turning on untilled areas in a configuration that allows all of these turn spots to be tilled in the final pass out of the area. Another adjustment in dry periods is to shorten the time between final tillage and planting. In extremely dry conditions, I have both the planter and the tillage unit in the field at the same time so that I can till and then plant small strips alternately. Another possible strategy is to plant deeper to place the seeds in more moist soil. The disadvantage here is that emergence is slowed.

It might seem that following these suggestions about moisture management will leave only a narrow window of opportunity for planting. However, this is not the case. Once fields dry after even a modest rain, there are usually sufficient reserves of moisture for planting for as much as two weeks or more with careful management.

Notice how these guidelines coincide with other agronomic reasons not to plant in soil that is too wet or too dry. They would be useful regardless of whether one used chemical or nonchemical weed control. Soybean seed, for example, planted in soil too dry to sprout is susceptible to shriveling and losing its germination. Conversely, there are many notorious problems associated with working and planting in wet soil.

Much of this management of soil moisture comes to little when it rains immediately after planting. For then weeds can germinate as the crop does. However, it often does not, and for every hour without rain, the crop has that much of a head start over weeds. In the final analysis, the most that can be expected of farm planning is that it improves our advantages for dealing with uncertainties.

MANAGING SURFACE RESIDUE

Another consideration at planting time is the high volume of residue remaining on the surface as a result of minimum tillage. Such conditions can generally be handled quite easily with a well-designed modern planter which is adjusted appropriately. The Case IH Early Riser and the John Deere Maxemerge both have sharp, staggered disks for furrow openers, a necessary feature for working well in heavy residue. Occasionally stalks or roots will lodge at the furrow closing wheels, pushing up soil and eventually plugging the flow of seed. To deal with this problem, the Case IH 900 series planter has been modified with staggered furrow closing wheels. I adjust the furrow opening disks so as to make as little furrow as possible. Smaller furrows help to prevent furrow soil from being dragged over the row during preemergence harrowing, a situation which would lengthen emergence time.

SEEDING RATE

Still another planting consideration in a pesticide-free system is the rate of seed sown. In general, seed population for all row crops should be increased slightly. One reason is that higher rates partially compensate for the thinning effects of rotary hoeing and harrowing. In addition, thicker populations shade the ground more quickly, thus contributing to weed control. I commonly increase the planting rate by 5 to 10%.

Mechanical Weed Management

So far in this chapter I have discussed three important components of weed control: crop rotation, tillage, and strategies for planting the crop. I turn now to the fourth aspect of active weed control in row crops: postplanting destruction of weeds by rotary hoeing, harrowing, and cultivation.

ROTARY HOEING

Though all features of the weed control system which I am describing can contribute to its success, the one that I consider essential is effective rotary hoeing of row crops. A large rotary hoe plays a crucial role in reducing and eliminating herbicide use. Because of this importance, I will explain in detail how to select an appropriate machine, how to use it effectively, and how to time the hoeing operations for best weed destruction. In doing so, I will also give attention to making rotary hoeing compatible with conservation structures and procedures, such as terraces, waterways, contour cropping, and minimum tillage.

In selecting a rotary hoe, keep these considerations in mind. First, I recommend choosing the widest hoe that fits a budget, terrain, and horsepower. Although 90 horsepower tractors are commonly used to pull 14- and 16-foot hoes, a tractor that size is capable of pulling a 20- to 30-foot hoe safely and effectively. Larger hoes provide several important advantages. One is timeliness. Because rotary hoeing is central to this weed control strategy and because timing is crucial to its effectiveness, one wants to be able to cover as many acres as quickly as possible when the time is right, especially, say, if rain is threatening. A second advantage of a larger hoe is that it potentially minimizes compaction. A larger hoe spreads the weight of the machine over a wider area and creates fewer wheel tracks, the place where its performance is poorest. A third reason is that a larger hoe reduces the total number of turns in the turn rows. Because a field will sometimes be hoed as many as three times in a season, it is important to minimize damage to the crop from turning. And finally, a larger hoe is easier to hold in place on the back slopes of steep terraces. That is because the wider the hoe, the more likely the tractor will not be on the steepest part of the back slope.

Another important feature is folding wings, with gauge wheels and separate hydraulics for each wing. For example, the popular 21-foot hoe can usually be purchased with a rigid tool bar or with a tool bar consisting of a 12-foot center and two folding wings. One obvious advantage of the folding model is convenience of transport. Long hoes without wings can be moved on the road only by unhitching and then hooking up to a transport hitch at the end of the hoe. Another advantage of the folding-wing models is that they fit a wider variety of field configurations. For example, in terraced or contoured fields, some unhoed portions of the point row area will be narrower than the hoe. Delicate small plants should usually not be hoed a second time. The problem can be dealt with by raising the wings and hoeing with only the center section. A final advantage is that a winged hoe equipped with gauge wheels can flex to conform to contours and terraces, thus allowing all of the hoe to function as intended. This feature is a virtual necessity in hilly, terraced fields.

Any hoe, regardless of size, will serve its purposes better if it has the following additional features. First, it should be a model which is mounted on the two- or three-point hitch rather than a pull type. The advantages of being able to raise the implement completely are that it facilitates transport, permits direct backing, reduces damage to turn rows, and makes unclogging the hoe easier. Second, the hoe should either be a model designed to deal with heavy residue or should have trash deflectors. These are bands of steel on the mounting arms which prevent trash from lodging between a mounting arm and the hoe wheel. If a wheel is stuck while travelling over a planted row, it tears out a portion of it. Trash deflectors are especially important to efficient hoe operation in reduced tillage situations. Third, the wheel bearings should be mounted with bolts rather than rivets so that they can be changed more easily in the field. Fourth, the hoe can be accommodated to a wider variety of field conditions if it is a model which allows the pressure on hoe wheels to be adjusted. Pressure can be varied from about 15 to 50 pounds per wheel on some models. And finally, the width of the hoe must match row spacing. For example, a 160-inch hoe, which is designed primarily for four 40-inch rows, cannot be used on 30-inch rows without either overlapping a row or leaving a row unhoed.

Just as important as selecting an appropriate rotary hoe is using it correctly. In this section I describe procedures for operating a larger model (21 to 31 feet wide), but many will also apply to the use of smaller hoes.

There are a number of operating procedures and cautions which, if followed, will make a large hoe more manageable and extend its life. One imperative is to *never* begin turning—or even turn the front wheels of the tractor sharply after stopping—until the hoe has been raised off the ground. Turning with the machine in the ground can cause bent or broken teeth, stressed wheel bearings, and bent mounting arms. Second, consider backing around turns. This reduces the amount of effort needed to maneuver a large hoe through the turn, and it reduces the brake and front tire wear associated with turning. When backing around, the weight of the hoe is partially pushed towards the tractor. Third, if the hoe is very wide (28 feet or more), turns in nonlevel areas may gouge terraces, waterway edges, and humps, with the possibility of damage to the hoe ends. This problem can be avoided by routinely raising the wings approximately 5% before each turn. However, to avoid damaging the inner wheels of the wings, the wings must be raised and lowered while the center section of the hoe is in raised position. And finally, a very wide hoe in raised position may rock against its mounting arms if it is driven into a depression, such as a waterway or terrace channel, at an angle. The resulting stress on the rear end of the tractor can be limited by slowing down, by driving straight into the depression, or, if the hoe is already rocking, by gently setting it down to stop the motion.

There are also a number of procedures and considerations in operating the rotary hoe that contribute to its effectiveness in destroying weeds. First, hoeing is most effective in the hottest part of the day, ideally in hot, sunny, windy weather. Evening hoeing is less effective. Thus the importance of spreading out planting dates, for if too many acres are at the same stage, the opportunities for timely hoeing are greatly reduced. Second, it is best to do turn rows first so that they are hoed before that portion of the field is packed by the repeated turning of the equipment. Third, hoeing is more effective at higher speeds. This is because the hoe teeth, which are shaped like little spades, do their work by penetrating the soil and flicking it backwards as the wheel rolls along. Popular speeds range from 7 mph on very small, delicate crops to 13 mph on better established crops. Fourth, given the importance of speed, the hoe must be started quickly and stopped abruptly to ensure that the ends of fields are hoed effectively. When possible, develop speed before coming to the unhoed area.

A special challenge to operating a rotary hoe efficiently is presented by heavy corn and milo residue. With corn residue, the problem is that roots lodge between the wheel and support arm of modern, single-row-style hoes. Older-style hoes with rows of wheels on axles have very few plugging problems. A partial remedy for me has been to add trash deflectors (about $250 on a 21-foot hoe). Another partial remedy is speed. In trashy conditions, I try to operate at 10 mph or faster. With milo residue, plugging has not been a serious problem except when the residue is mixed with morning glory vines. Of more concern is that there will be so much residue that the hoe cannot penetrate it well. However, this has not been a chronic problem and has never caused me to lose overall weed control. In general, because rotary hoeing shreds residue aggressively, there will be few problems with subsequent hoeings. This probably also explains why I have had virtually no problems with cultivating row crops in fields with heavy residue.

There is one important caution about operating a rotary hoe in trashy conditions. When the machine plugs, it will be tempting to partially raise it and then back up to unplug it. Absolutely resist this temptation, for the procedure is likely to break mounting arms.

In addition to selecting an appropriate rotary hoe and operating it effectively, the timing of the hoeing in relation to the growth of the crop and the weeds is also essential. My experience has been with using the implement on corn, soybeans, grain sorghum, and, more recently, wheat.

Row crops may be successfully hoed prior to emergence. Corn and sorghum may be hoed until the spike stage breaks into the first two leaves. Beans may be hoed until the hook is about to break through the soil. In each case, wait three days or so after planting to permit weeds to germinate. The problem with preemergence hoeing is that it often commits the farmer to three hoeings. The crop will most likely require a second hoeing shortly after emergence and then a third hoeing seven to nine days later. By the time for the third one, the soil may have become too compacted for maximally effective hoeing. Nevertheless, sometimes preemergence hoeing is advisable, especially if rain threatens.

Corn and soybeans can withstand hoeing three or four days after emergence, especially at slower speeds. The complication at this stage is more often covering up the crop than tearing it out. One thing to do is test hoe a strip and then count to see exactly what the loss is. Seldom is it as bad as it looks from the tractor seat, and increased planting rates can largely compensate for the damage. However, if the field is on a hillside or if ridges were created during planting, the covering problem may be too severe. Sorghum is more susceptible to damage just after emergence. Therefore, it should be about three inches high—requiring as long as a week after emergence—before it is hoed. Because of this need to wait longer before hoeing sorghum, there is more incentive to do a preemergence hoeing than there is for corn or soybeans. Usually all three row crops will require a second postemergence hoeing. All can withstand a fast, aggressive hoeing up to a height of six inches or so. However, as plants grow larger, losses on turn rows increase.

I have recently begun to rotary hoe wheat in the fall to destroy pennycress. I have observed almost no negative impact upon the wheat; even tire tracks do not appear to be a problem. The only concern is the possibility of covering up wheat on terrace back slopes. I have left the steepest ones unhoed in wheat that was about four inches tall.

HARROWING

A spring-tine harrow nicely complements a rotary hoe in a weed control strategy for row crops. While it is not as versatile, it is uniquely suited to certain situations. For example, it is possibly the only implement suited for use in a sometimes crucial but brief stage of weed control in soybeans. Its main limitation is that it does not work well in fields with crop residues. However, in a diversified

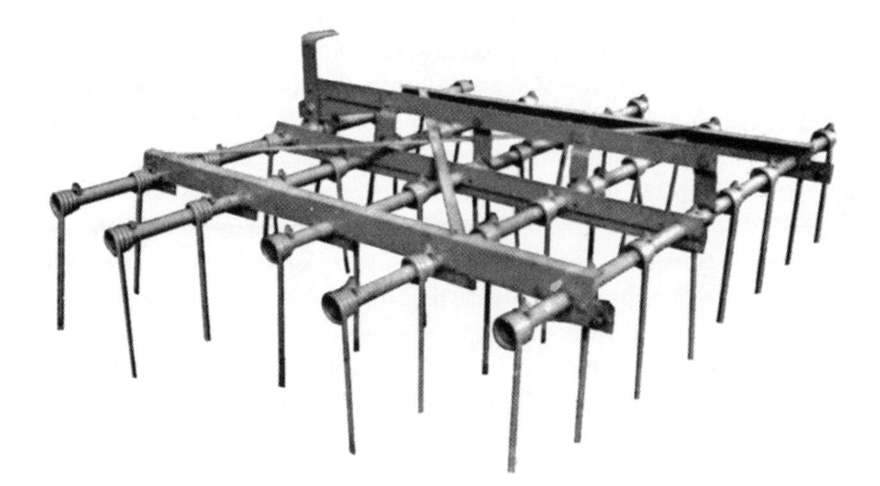

Figure 3. Section of a spring-tine harrow. Courtesy of Jack Kovar Sales Co.

cropping system there are often fields without much trash; therefore, the spring-tine harrow can have a useful role. I will describe how to select an appropriate model, how to use it effectively, and how to determine what situations to use it in.

A harrow for weed control in row crops is an inexpensive implement which differs greatly from spike-tooth drag harrows. For one thing, it has a fully mounted tool bar. Extending back from the tool bar are booms from which are suspended several independent sections. Each section consists of configurations of spring-loaded tines which are something like the tines of harrows mounted at the rear of tillage equipment. The crucial difference is that the tines are five-sixteenths of an inch in diameter, which is one-sixteenth of an inch smaller than those on tillage harrows. This difference allows the tines to achieve the correct action in row crops. The tines are also long—about 15 inches. This keeps the framework of the sections up out of taller crops and also extends the useable life of the tines. The sections also have adjustment arms which allow the pitch of the tines to be varied. (See figure 3 for a picture of a harrow section.)

The best width for a harrow depends on the farm terrain. On flat land, one consideration is the potential that the ends of the harrow will gouge during turns. Unlike rotary hoes, the harrow's tool bar wings usually cannot be raised slightly to avoid this problem. Another consideration is convenience of transport. If a greater portion of the implement's width is in the center section, there may be more problems going from field to field. But if a substantial portion of the width is in the wings, there may be complications in raising them manually for transport, perhaps requiring the installation of a winch. These considerations suggest that it would be quite difficult to efficiently use a harrow much wider than 35 feet on flat land.

For a farm with terraces, a shorter harrow will probably be required to function correctly in the terrace channels. These areas can be harrowed effectively by lowering the tool bar until the ends are just above the ground. This creates slack in the chains that support the middle section, allowing it to reach into the depression and function properly. Based on my experience, the necessary balancing for this technique can be achieved with a harrow no wider than 26 feet—that is, one set to cover ten 30-inch rows or eight 38-inch rows.

As with a rotary hoe, a harrow must be matched to a certain number of rows. Choosing one which is the wrong width may necessitate working a row a second time on the return pass in order to cover all areas. This would be an intolerable situation.

Operating a spring-tine harrow in row crops requires skill and care. There are several techniques and considerations which can contribute to success. One is that the sections must be suspended correctly in relation to the tool bar. The tool bar should be lowered so that there is modest slack in each of the four support chains that hold each section. If the tool bar is too low, the sections will ride on the front portion, causing them to lurch and perform poorly. The aggressiveness of the harrow's action can be adjusted in three ways. One is the pitch of the tines. Another is weight. A simple way to add weight is to place a tire on each section. The third adjustment is speed. Going too fast can result in covering up all but the largest crops.

Although, as I mentioned, the spring-tine harrow does not work well in heavy residue, there will be situations with moderate residue when it will be desirable to use it. Some useful strategies for harrowing more effectively in moderate trash are to select a flatter tine pitch, to work in the driest conditions (for example, afternoon rather than morning), and to stop occasionally and lift the entire harrow. That is a key reason these harrows are fully mounted, a feature that also helps to reduce crop stress when turning around.

It is important to avoid wet fields with a harrow, as it is with a rotary hoe. For one thing, soil will be compacted. Furthermore, the harrowing will be much more effective in drier soil. Another caution in operating the harrow is to never back up with it on the ground. Nor should it be turned so sharply on a tilt that the outer tines are pulled backwards. In both cases the tines are likely to break.

One use of the harrow is before the crop has emerged, timing it as I described earlier for preemergence use of the rotary hoe. Though I am not aware of any particular advantage of the harrow over the hoe at this stage of the crop, at least it is available if the hoe is broken down or if there is a reason to have both machines operating at the same time. With corn and sorghum, the action of the implement mainly bends the spikes of the plants over. Even if they break off, there will be regrowth. If preemergence harrowing is a possibility, it is important to avoid creating ridges during planting. The harrowing may then put the crop too deep, delaying emergence.

Postemergence harrowing differs in several ways from postemergence rotary hoeing. One is that sorghum is too fragile in its early stages to tolerate harrowing. Another is that corn and soybeans must be allowed to grow a bit more before they are harrowed, since harrowing is generally a more aggressive procedure. On the positive side, a harrow can be used on corn and soybeans at a much larger stage of growth. Corn can be harrowed up to 12 inches in height with virtually no damage, thus providing one more opportunity to cultivate directly in the row rather than being forced to use a shovel cultivator.

Perhaps the most important use for the harrow is that brief but crucial stage when soybeans are too big to rotary hoe but too small to cultivate. At this point the hoe does too much damage, including shredding leaves and sometimes pushing the plants over. And a shovel cultivator is not good at this stage because the foliage is still too low to permit the shovels to push soil to the stalk. That harrowing permits another cultivation in the row is especially important if there has been rain after planting. It provides a good opportunity to dislodge late-sprouting weeds such as velvetleaf and pigweed. At this stage, harrowing has little effect on population. However, the operator must be careful not to drive too fast, because even larger soybean plants remain vulnerable to covering. This particular harrowing of soybeans has, for instance, made the difference between excellent weed control on one field and a less satisfactory situation on an adjacent one where trash prevented using the harrow.

CULTIVATING

The final technique of mechanical weed control which I will discuss is shovel cultivation. The main point I will stress is that it should begin as soon as possible. The criterion should not be, as for starting hoeing, when weeds are sighted, but rather when the crop plants are high enough so that soil can be moved to them without covering their lower leaves. There are three reasons for this urgency. One is to hedge against a dry spell. If the weather turns dry, soil that has been driven on in hoeing and harrowing will become increasingly hard, even to the point that a shovel cultivator cannot function properly. Another reason for cultivating sooner is to hedge against an extended wet spell. Early cultivation reduces the chances that lingering rains will wreck the weed control program. A field planted to row crops without herbicides is not under weed control until it has received its first shovel cultivation. The third reason to cultivate sooner is to quickly get at deeply rooted weeds. For example, cocklebur is notorious for its capacity to elude cultivator shovels. An early start against weeds such as this improves chances of uprooting them and of covering weed plants in the row.[8]

My cultivator is an older model with rigid shanks rather than spring-loaded shanks as on a field cultivator. I find that rigid shanks provide better penetration in hard soil. I use wide rear sweeps for aggressive action against weed roots. And I always drive fast enough to push soil into the row. Unless conditions have been rainy, I find that a second shovel cultivation is rarely needed.

Replanting

The procedures and strategies which I have been describing are usually effective. Results will be influenced, however, by a myriad of variables, some of which the farmer has little or no control over. A thorough understanding of and commitment to these weed control procedures is no guarantee of success. Non-herbicide weed control strategies are not sophisticated enough to cope with the ultimate nightmare: an extended wet spell in June, especially one beginning just after soybeans and milo are planted. Although staggered planting will minimize the impact of this weather misfortune, sometimes a drastic response will be required. That is to replant fields which are out of control.

One can tell within 10 days after emergence whether the crop is in or out of weed control. If it is out of control, shovel cultivating will not reestablish control. Faced with a field with weed problems after completion of rotary hoeing, a farmer should be prepared to replant. Being genuinely prepared involves several considerations: sufficient financial resources, machinery for doing the work expeditiously, and psychological readiness. In the absence of any of these, a farmer is apt to be less inclined to replant.

There are a number of considerations and strategies that will make the possibility of replanting more bearable. For one thing, large, modern machinery will minimize the time required to get a new planting into the ground. One way to help speed the process is to delay for as long as possible unhooking the planter and returning it to storage. Furthermore, farmers who save and condition some of their own soybeans for seed can prepare and keep on hand an extra quantity of a short-season variety at little additional cost. At least, a source of an appropriate seed should located ahead of time. Also easing the financial pain is the fact that many seed companies offer significant discounts, especially on corn and sorghum, when replanting is necessary. Sometimes it is a complete replacement policy.

The possibility of needing to replant is not unique to pesticide-free farming and should not be considered a special obstacle to nonherbicide weed control. Herbicides sometimes fail to perform, creating distinct problems for replanting. First, at the time of replanting, the farmer will already have invested the costs of the herbicides, whereas the farmer using nonchemical means may—but then also may not—have already incurred weed control costs. Second, replanting options are reduced because of chemical incompatibilities among crops. Third, the failure of the herbicide to perform creates uncertainty about whether enough herbicide remains or whether a second application would be too much.

Weed Management Progress and Problems

I will now describe the state of control which I have achieved for specific problem weeds and present some methods for dealing with a particularly troublesome one.

Several weeds that are major problems in this region are generally under excellent control on my farm. There is virtually no shattercane except what is present in the sorghum seed we plant. Control seems to be the result of rotating the sorghum grains with other crops, especially alfalfa, wheat, and oats. In fields of milo we do hand roguing, either dropping the heads on the ground or, if the seed might be viable, removing the heads from the field. Hemp dogbane was a rather severe problem at the beginning in the 1970s. However, crop rotation incorporating alfalfa as well as moldboard plowing in the hottest, driest part of summer quickly arrested it. Foxtail and other grasses are no longer a concern. This exceptionally good control probably results from the effects of crop rotation and the susceptibility of these weeds to rotary hoeing. When I began operating the farm in 1975, common milkweed (*Asclepias syriaca* L.) seriously infested several fields. However, we quickly obtained excellent control through diversified cropping and tillage, especially moldboard plowing.

In the case of several other potentially bothersome weeds, we have achieved moderate control. Velvetleaf, pigweed, and lambsquarter (*Chenopodium album* L.) are opportunistic weeds on the farm. Control is excellent when conditions permit good timing of weed control procedures. However, wet weather can reduce that control. The problem with controlling cocklebur is the limited effectiveness of rotary hoeing because of its deep germination and the difficulty of uprooting it with shovel cultivation. Measures for dealing with cocklebur include hand roguing, late planting, which allows several crops of burs to be destroyed, and seeding infested fields to small grains, which leaves the weed vulnerable to tillage in July. Morning glory remains a problem in a few fields but has diminished in others for reasons that I do not yet understand. I am still looking for the crop and tillage sequence that will provide predictable control.

And finally, two weeds that I have under only fair control on my farm are bindweed and pennycress. Pennycress is sometimes a challenge in alfalfa, wheat, and oats, but not in row crops. The infestation is probably due to a high percentage of sod-base crops in the conversion years, crops which promote pennycress. This situation is being brought into balance with a return to a higher percentage of row crops, crops which interrupt the pennycress life cycle. I am also optimistic about the newly adopted practice of rotary hoeing wheat for pennycress.

Bindweed deserves special attention. I will discuss it in detail because of its extraordinary tenacity and because of my extensive experience trying to deal with it. I know of no crop rotation, no single tillage procedure, no single chemical application, no allelopathy situation, and no insect process that will eradicate an infestation. Thus, bindweed is a very serious problem.[9]

Bindweed is a perennial noxious weed of the morning glory family. A number of its growth characteristics contribute to making it especially tenacious. These are summarized by Martin, Leonard, and Stamp (1976) as follows: "This plant . . . is a perennial that propagates by seeds and by lateral creeping roots from which rhizomes and then new stems and roots arise. The roots may grow 1 inch a day and penetrate to a depth of 12 to 20 feet or even more. The lateral roots, which can spread several feet in a year, are found mostly at a depth of 12 to 30 inches. The stems are smooth, slender, slightly angled vines that spread over the ground or twine around and climb any erect crop plant. The vines may be 1 to 6 feet in length" (312).

Several nonchemical procedures have been developed to deal with bindweed. My own method is to gradually starve the root system by isolating infested areas, removing them from crop production, and doing continuous tillage on them. For this, I prefer a chisel plow with wide sweeps—at least a four-inch overlap. I try to work at a depth of three inches and to operate with the implement level. To insure thorough disruption of the roots, I often go over a plot two or three times. If I have a different tillage implement attached to the tractor when the time comes to work these infestations, I use that machine instead. These include a moldboard plow, an offset or a tandem disk, and a field cultivator. When I use a field cultivator, I am careful to remove all vines and root pieces from shanks and tines before using the implement in another field.

In order to deplete the roots early in the season, I begin tillage when the plants are a few inches high. Then I till approximately every two weeks for the rest of the summer. The length of the interval depends on how long the patch has been undergoing tillage; those just beginning treatment grow back most vigorously. It is best to let the plants grow back to a few inches before the next tillage.

Advantages of this method of dealing with bindweed are that it isolates the infestation, thus preventing spreading, and it eventually depletes the roots and the seed base. However, there are also problems associated with it. One is that timely tillage may be impossible—thus eroding the previous gains—because of other pressing farm work, extended wet spells, and very dry periods, which on my farm make the silty clay soil too hard for effective tilling. Another is that the exposed soil may be vulnerable to erosion. A third drawback is that the area being treated is out of crop production. And finally, the procedure cannot be used on infestations in untillable areas, such as grass waterways, fence rows, and orchards.

Some of these disadvantages can be at least partially overcome. To minimize winter soil loss, I have begun to seed the areas to rye. This provides protective cover and pasture, and it may have a smothering and allelopathic effect on the bindweed. And, of course, erosion is minimized by terraces and by the fact that the patches being treated will have growing crops around them. For infestations in untillable areas, burning is a possible control measure. According to Hanson, Keim, and Gross (1943), "Eradication can in most cases be accomplished in two to three years by . . . burning the infested area every 10 days. A light searing of the foliage is just as effective as reducing the vegetation to ashes. It is very important that the entire area be given a thorough and uniform searing so that there are no plants skipped" (20). A 1923 Kansas Experiment Station article asserts that, because of remaining seed, areas where the weed has been killed should not be planted to small grains for several years (Call and Getty 1923, 12).

A method of bindweed control using tillage followed by a smother crop is described by Call and Getty (1923, 17) and Martin, Leonard, and Stamp (1976, 307). First, continuous tillage much as I have just explained is carried out for one full season and until July 1 of the next. At that time a smother crop such as sorghum, Sudan grass, winter rye, or wheat is planted on the infested area. The sources suggest that the procedure should be followed for several years. I have not been able to try this modification because, as on a typical organic farm, July 1 is virtually the busiest time of the year. This method may work well on farms with more help available.

The two methods of bindweed control just described are best suited for smaller patches. For intermittent infestations throughout an entire field, farmers may wish for an effective method which includes some crop production. In a Nebraska Experiment Station article, Hanson, Keim, and Gross (1943, 12–13) present a four- to six-year procedure which includes some forage and/or pasture and some wheat or rye for harvest. They schematize their plan as shown in Table 3.

Another eradication strategy that also includes some crop production is presented in Kirschenmann (1992, 17). It is a five-year plan developed for the northern plains.

Recently a method of bindweed control has been devised by Mattingly (1985, 35) which draws upon the allelopathy of pumpkins and banana squash. He plows in May and then in June seeds the pumpkins or squash a foot apart in 60-inch rows. According to Mattingly, the vines shade out and strangle the bindweed, preventing it from making seed. After harvesting the pumpkins or squash, he disks the vines in the fall. Mattingly reports no recurring bindweed for nine years. In 1991 and 1992 I tried this procedure and had excellent results. The next step is monitoring the plots for recurrence of bindweed in 1993.

Another recent suggestion is that certain insects offer possibilities for battling bindweed. Among them are the Argus tortoise beetle, the moth *Tyta luctuosa,* the mite *Aceria malherbae,* the flea beetle, and seed beetles (Rosenthal, Andres, and Huffaker 1983, 18). However, there are at least two problems with this approach. First, most of these insects also feed on such plants as sweet potatoes. And, second, the basic strategy is control rather than eradication. It remains to be seen whether this is satisfactory, given how easily bindweed is spread by tillage.

Table 3. Procedure for continuous cultivation[1] for eradication of bindweed

Jan.	Feb.	Mar.	Apr.	May	June	July	Aug.	Sept.	Oct.	Nov.	Dec.
1ST YEAR											
Cover crop or crop residue --------				XX[2]	XX	XX	XX	XX	Rye or wheat for winter cover ---		
2ND YEAR											
Crop residue, pasture, or hay --------				X[3]	XX	X	XX	X[4]	Wheat or rye for seedling control ---		
3RD YEAR											
Wheat or rye for grain and seedling control --------------						Cultivate immediately after harvest to control seedlings ---------					
4TH YEAR											
Repeat the 3rd year procedure for the 4th and possibly for the 5th and 6th years to destroy seedlings.											

[1] Each cultivation is represented by the symbol x.

[2] Begin duckfooting or subsurface tillage about three weeks after first emergence of bindweed (May 5 to 25) and continue at 2-week intervals until seeding time or until the end of the season on small areas where no cover crop is planted.

[3] Begin at the same time as in the first year and either cultivate the cover crop with a subsurface tiller to leave residue on the surface of soil that is subject to erosion, or remove the crop by pasturing or cutting for hay where the land is not susceptible to erosion, and cultivate with an ordinary duckfoot. Where hay is to be removed, it may be necessary to delay beginning cultivation for a week or 10 days for the crop to be cut at heading time. Because of the density of the crop at this period, there will be little gain made by the bindweed, but the area should be cultivated as soon as the hay has been removed. Cultivate every three weeks or eight days after re-emergence of the bindweed for the remainder of the time necessary for eradication.

[4] Bindweed is usually eradicated by the end of the second year with this method.

Source: Hanson, Keim, and Gross 1943, 12–13.

Conclusion

The complexity of pesticide-free weed control is undeniable. It emerges out of a context of sophisticated crop rotation. From that, aspects of it commence in early spring and continue through much of the summer. Is all this complexity a grave problem for alternative agriculture? Not necessarily. There is no reason why mastering the effective operation of a rotary hoe, for example, need be any more arduous than learning to calibrate pesticide sprayer nozzles. More generally, alternative farmers widely proclaim that management is not burdensome and becomes easier each year. A much closer look at convenience, and other points of contrast between alternative and conventional systems, follows in chapter 5. We turn next, however, to a look at the role of livestock in alternative farming.

4

• • •

Livestock

Some readers will be surprised and a few even dismayed by the importance which I place on livestock in sustainable farming. Therefore, I will begin this chapter with a detailed explanation of why livestock is a crucial component of a generally applicable system of alternative agriculture. Next, I address a central criticism of alternative agriculture: that it requires too much livestock. Following that is one suggestion in the controversy about livestock and the environment. I conclude with two very practical matters.

In my discussion I use beef cattle for illustrative examples both because that is what my experience is with and because ruminants are ideally suited to be the livestock constituent of pesticide-free farming. And I use organic farming as the basis of analysis, thereby accepting the greater challenge of defending what is often regarded as the least plausible alternative system in order to make the strongest possible case for alternative agriculture in general.

The Question of the Indispensability of Livestock to Organic Farming

How important is the question of whether livestock is indispensable to organic farming? As it turns out, very important indeed. The structure of livestock production has proved remarkably sensitive to features of federal policy, especially tax policy. For example, components of capital gains, cash accounting, investment credit, and accelerated depreciation, especially prior to 1986, have contributed to increased size of nonfarm investment in and confinement of livestock operations.[1] The relevant consequence is reduced integration of livestock into farming systems. If organic agriculture is important and if the careful integration of livestock into those systems is vital, then policymakers would have powerful incentive to avoid policies that inappropriately separate livestock from farms.

An animal welfare movement gains support; a subgroup of that movement is opposed to livestock.[2] Once again, suppose organic farming is essential to an emerging concept of appropriate agriculture, and suppose livestock is essential to organic farming. In that case, livestock would have an importance to the future of agriculture which must factor into discussion of the moral acceptability of livestock in general.

A third reason stems from resistance to livestock production from certain perspectives about land and resource management. In this view, livestock production, at least in its current configuration, is not consistent with coming demands of resource efficiency (Robbins 1987). It focuses, for example, on accelerated energy consumption and land degradation associated with ill-conceived livestock systems. If organic farming is the appropriate response to resource concerns and if livestock is a necessary constituent, then criticism of livestock from this perspective would be modified.

Finally, as farmers become attracted to organic systems, it would be advantageous to be specific about the requirements of those systems. We would not want to be led to believe that organic systems can be viable without livestock if that is false. There is ample reason to try to think precisely about the relationship between livestock and organic farming. The answer bears on individual decisions about farming systems and larger ethical and policy decisions about the future of agriculture.

Livestock raising is important to organic farming because it contributes crucially to achieving four central objectives:[3]

- Soil conservation. Livestock give more economic viability to the crop rotations, the conservation structures, and the grasslands required for effective conservation.
- Pesticide elimination. Livestock are key to the rotations and legumes that make farming without chemicals possible. They consume the forages of typical rotations and rid grain fields of grain that could grow as weeds in the next year's crop.
- Financial stability. Livestock are a key component in a mutually reinforcing system which creates stability through diversity, risk management, and making marginal land productive.[4]
- Nutrient recycling. Livestock play a crucial role in establishing and main-

taining soil fertility, including such important components as nitrogen, phosphorus, potash, trace minerals, humus, and microbial life.

An obvious and legitimate question is: Could these objectives for organic farming be achieved without livestock? The initial answer is: Yes, it is possible to imagine such farms, and they no doubt exist. For example, fragile land could be protected by putting it in a speciality crop such as Christmas trees or nut trees or perhaps by using it for commercial recreation. The rotation crops necessary for eliminating pesticides could be sold to farmers with livestock and the fields could be grazed by neighbors' livestock. To attain some measure of financial stability, perhaps organic farmers could compensate for the absence of livestock income with special markets for their other crops. To recycle nutrients, an organic farmer might import manure from a nearby livestock operation, municipal sludge (if organically approved), or paunch manure from a nearby packing house.

No doubt there are organic farmers who have used these and similar procedures to create viable operations without livestock. They are to be commended for taking advantage of the resources and possibilities available to them. However, notice that their success in meeting the objectives of organic farming depends either on someone nearby having livestock or on a special situation which would not be available to all, or even most, farmers. The dispensability of livestock to organic systems in general is not established by relying upon someone else's livestock or by taking advantage of special markets, proximity to population centers, or speciality products with limited demand. It does not follow from the fact that some organic farmers are successful without livestock that all or even most organic farms could be. The significant and realistic question is whether, without livestock, organic farming could be widely and generally applicable.

The question of farming organically without livestock should be posed again, but with certain restrictions. That is, what would an organic farm be like

- if there were no livestock component and no available agronomic benefits from livestock owned by others (imagine an agricultural setting that is a culmination of current trends, a setting in which all livestock is maximally concentrated in just a few locations, resulting in a continuing market for feed grains, but leaving livestock almost entirely remote from farms)
- and if the farm did not benefit from special soil amendments or markets not available to organic farmers generally?

We can now examine how such a hypothetical organic farm would cope with the four considerations for having livestock. It will be referred to as the generalized organic farm without livestock.

The impediments to adopting appropriate soil conservation measures would primarily be economic. Quite likely the farm requires some grassed waterways, and there may well be some marginal land which is not suited to tillage and should be in grass. Without livestock, it will be very difficult or impossible to utilize the grass from these areas. Similarly, good soil conservation requires a crop rotation which includes forage crops and small grains. However, these are generally not very profitable unless they are used on the farm as resources in a livestock operation.[5]

Second, the absence of livestock would also make it much more difficult to farm without chemicals. For one thing, the diversified crop mix and rotations necessary for dealing with weeds and insects would be less feasible without livestock. In addition, alfalfa, which is virtually unique in its capacities for building soil tilth, suppressing weeds, and fixing nitrogen, would not be economically viable at all in a generalized organic farm without livestock. Moreover, the problem of volunteer plants in the next year's crop would have to be dealt with either by adjusting the rotation or by hand roguing. Thus, without livestock, the goal of farming without chemicals is much more difficult to attain and sustain.

Third, the absence of livestock from a generalized organic farm creates several problems for financial stability. Most obviously, there will be no income from livestock. Also lost will be the capacity of livestock to salvage damaged crops, a capability which can significantly lessen the financial impact of a disaster. Furthermore, without livestock, there is virtually no way to generate income from marginal lands in a responsible manner. Thus, while we cannot conclude that a generalized organic farm without livestock will be unprofitable, it clearly would be much more vulnerable to weather and market conditions.

Finally, the most severe challenge to a generalized organic farm without livestock is posed by the objective of nutrient recycling.[6] Livestock recycle so many soil amendments that the subject is very complex. To illustrate, I will concentrate on just one—phosphorus. According to Goldstein (1986, 51), it is often the limiting factor in soil fertility. The problems are that only a small amount is present in soils, that much native phosphorus is unavailable to plants, and that added phosphorus is quickly unavailable (Brady 1974).

Organic farmers have traditionally—and wisely—used animal manure to supply phosphorus. One reason is that some of the phosphorus is immediately available to plants. A second reason is that livestock manure enhances the availability of inorganic soil phosphorus by affecting positively four—and perhaps all five—of the factors which control it. These are soil pH; soluble iron, aluminum, and manganese; iron-, aluminum-, and manganese-containing minerals; available calcium and calcium minerals; and activities of microorganisms (Brady 1974, 460).

Consider, then, the problem posed by the concept of a generalized organic farm without livestock. By definition, it cannot address the phosphorus problem with livestock manure or with other amendments not available to organic farming in general. What remains to be used, it appears, is rock phosphate. Unfortunately, its phosphorus is substantially unavailable to plants (Lengnick and King 1986, 108).

Basic aspects of organic systems—soil conservation, nonchemical methods, financial strength, and nutrient recycling—are compromised without livestock. Thus, the conclusion seems inevitable that, at best, organic farming is made vastly more difficult and may not be workable at all without livestock.

The Question of Whether Organic Farming Requires Too Much Livestock

A major objection to organic agriculture is the following: If every farmer practiced it, our country would be overrun with livestock—much more than we need or want. One example of this assertion which makes specific projections about nitrogen fertilizer is from the Potash and Phosphate Institute. Its Foundation for Agronomic Research of Atlanta Georgia states that "animal numbers (AANUS) would have to increase more than fivefold in the U.S., eightfold in the Corn Belt, and elevenfold in Iowa to supply enough manure N to replace fertilizer N." This is a serious challenge to organic farming, one which is sometimes taken as decisive in debates about the future of agriculture.[7] We need to examine it closely.

Testing actions or principles by examining what would happen if everybody did it—the generalization test—has a long tradition in ethical theory. And it appeals powerfully to our intuitions. However, for the results to be useful, the test must be applied to an accurate and fair description of the situation. For example, consider first what appears to be a straightforward application of the generalization test to livestock in agriculture:

> What if every farmer attempted to use manure to replace fertilizer at his or her current rate of use?

If this is the hypothetical test that organic farming must pass, it appears that organic farming fails. However, the question has three flaws. First, it assumes the cropping pattern of a conventional farm, one in which 50 to 100% of the acreage is in feed grains. Second, it assumes that manure would replace fertilizer at the rates currently used in conventional farming. Finally, it assumes that livestock manure is the only fertility alternative an organic farmer will wish to use. However, each of these assumptions is false, as we shall see.

Consider another formulation of the test, one which attempts to address the relationships among livestock, efficiency, conservation, and organic methods:

> What if every farmer were to raise livestock in a manner that permits efficient and conservation-oriented utilization of resources and that makes livestock and crop production mutually supportive?[8]

Framed thus, the test highlights a crucial problem with the way livestock are raised in conventional agriculture. The trend in our country is towards large-scale confinement rather than farm-based livestock production. However, this violates the stipulation that livestock production must be supportive of crop production. Here are two examples. The difficulty of efficiently utilizing manure from large-scale confinement lots violates the stipulation that livestock production be supportive of crop production. Among the obstacles to efficient manure utilization in large confinement lots are uncertainties about composition (salt, noxious weed seed, and chemical residue) and transportation. The second problem increases proportionately as feedlot size increases. The impossibility of permitting livestock in large-scale confinement to efficiently forage crop residue and grasslands violates the stipulation that crop production be supportive of efficient livestock production.[9]

The inappropriateness of large-scale livestock lots to organic farming illustrates the problem with the first version of the generalization test. It is not fair to evaluate the feasibility of organic farming by mapping bits and pieces of it onto bits and pieces of conventional agriculture. It is obvious that if we added the livestock necessary for generalized organic farming to the livestock currently produced in large-scale confinement the total would be excessive. But large-scale confinement production is not consistent with a central tenet of organic agriculture. The too-much-livestock criticism, therefore, expects organic agriculture to take responsibility for a component of conventional agriculture—livestock in large-scale confinement production—that organic proponents are entitled to repudiate.

Thus, we can begin answering the question of whether organic farming would require too much livestock by proposing that livestock production from large-scale confinement could be returned to farms without any increase in total numbers of livestock. This would provide a significant enhancement to organic agriculture. For example, in 1981 the number of beef cattle in large-scale confinement (1,000 head or more) in a 23-state feeding area was 16,862,000, or 73.3% of total production. The average population per lot was 7,524 head. Fed cattle from farms were only 6,152,000, or 26.7% of the total (Van Arsdall and Nelson 1983, 34). Thus, based on these figures, over 16 million head of beef cattle could be returned to farms and integrated into organic systems without increasing total numbers produced.

Still the question remains as to whether the number of cattle that could theoretically be shifted from feedlots to farms would be sufficient to meet the needs of organic agriculture. The argument, therefore, is not yet complete. I will attempt to demonstrate that the role of livestock in organic farming does not have to be as extensive as is sometimes suggested. I will do so by considering what amount of livestock is minimally sufficient to support an organic farm.

Earlier I described four salient relationships between livestock and the objectives of alternative agriculture: soil conservation, pesticide elimination, financial stability, and nutrient recycling. Let me now construct a model of a moderate-sized farm with a modest number of beef cattle in order to examine the extent to which the livestock would contribute to meeting these objectives. We will say that it is a 320-acre farm of the sort typical in the Corn Belt and Great Plains, that is, one suited to row crops but with a few acres which are steep, rough, or adjacent to creeks. The livestock operation is a 30-cow herd (essentially a one-bull herd), and calves are brought to finish on the farm. This would total about 71 head—that is, 30 cows, 1 bull, 27 calves (90% calf crop), and 13 yearlings (26 yearlings divided by 2 based on a 6-month finishing cycle).[10]

First, with respect to soil conservation, this number of cattle should be sufficient to promote ambitious conservation practices on 320 acres. If the land is typical, a few acres are best suited to pasture. In addition, there are—or should be—grassed waterways. The 30 cows will provide sufficient financial incentive to use the rough ground as pasture and to have the waterways from which to harvest grass. Even if the land is quite hilly, with 3 to 4% in waterways, the forage yield from about 13 acres—13 to 15 tons—is easily within the feed requirements of the cows and their calves. Conservation also entails crop rotation, typically wheat, oats, alfalfa, and other legumes, in addition to row crops. The ability of the cattle to pasture on the wheat and to consume oats and legumes makes the crop rotation much more feasible.

With respect to pesticide elimination, part of the strategy is diversification, rotation, and legumes, whose connection to cattle I have just described. Another role for cattle is grazing corn and milo residue to minimize volunteers in the succeeding year's crop. On most organic farms with diversified rotation, the percentage of acres in feed grains will be greatly reduced. Thus, 55 head of cattle (cows and offspring) should be able to clean up the prior year's crop residue on a 320-acre organic farm in normal winters.

With respect to financial stability, this size cow-calf operation will permit the sale of over 20 fat steers and heifers a year in addition to some culled cows. What is noteworthy are the opportunities to control costs of production. In this model, both the calves and almost all of the feed are raised on the farm. In addition, the operation is organized to maximize use of roughage and crop residue. And finally, this number of livestock would be adequate to cushion the financial blow from crop damage due to drought, hail, wind, early frost, and early snow. Even if it is impossible to harvest the grain, much of the plant and some of the grain would remain available for forage.

With respect to nutrient recycling, the livestock operation which we have established on this model farm would contribute significantly to meeting the fertility needs of the farm, but would not fill them completely. The nutrient recycling consists of the cattle eating roughage from pasture, waterway hay, legume hay, residue from fields, and grain. Grain production on this farm is likely to be 5,000 to 10,000 bushels. Suppose that 24 calves are fed 50 bushels of grain per head to complete their finish, a total of 1,200 bushels. This does not maximize nutrient recycling, as would feeding all the grain production of the farm. However, the opportunity to sell some of the feed grain crop does help to diversify income further.

It is also true that the manure from 30 cows and calves, a bull, and 20 to 30 calves in a feedlot is not enough to meet all the fertility requirements on 320 acres. According to Neuman (1977, 44), a 1000-pound beef cow produces 63 pounds of manure a day. Therefore, using the manure efficiently is a challenge, but not an insurmountable obstacle. On an organic farm, fertility results from the management of several variables in addition to livestock manure. These include:

- crop diversification which incorporates crops with lower fertility requirements,
- crop rotation,[11]
- legumes to fix nitrogen,
- alternative fertilizers, such as fish meal, green manure, and kelp, and
- yield goal adjustments to maximize efficiency.

Thus, this examination of a model farm with a 30-cow cow-calf operation puts the role of livestock in organic farming into perspective. The key points which emerge are that livestock have many important purposes in this kind of agriculture and that a modest number of livestock can fulfill most, even though not all, of these purposes.

What is "modest"? We can perhaps get a perspective on this by seeing how the number of livestock projected by the model organic farm compares to the number actually on farms in the county where I live. Cass County, Nebraska, is located along the Missouri River on the western edge of the dryland Corn Belt. In 1978 there were 333,183 acres in agricultural production, primarily corn and soybeans. The total number of cows, cattle on feed, hogs, and sheep in 1978 was 158,200 (Borchers et al. 1984, 4). That comes to slightly more than 2 acres per animal. Yet, for the model organic farm just described—that is, 320 acres with 71 total head of cattle—the ratio is slightly more than 4 acres per animal. The model, therefore, requires about half as many head of livestock per acre as were actually present in Cass County in 1978.

Granted, these calculations and comparisons do not answer precisely whether the amount of livestock required in organic systems is an appropriate amount over all. One difficulty is that Cass County, Nebraska, may not be representative. Another is that, to a certain extent, large animals are being compared with small ones. A more significant issue is that appropriateness is partially dependent on assumptions about how much meat should be consumed; those nutritional debates have not yet been resolved.

However, a more modest conclusion is possible. It is that the too-much-livestock criticism is without foundation—that is, the charge that organic farming, if generally applied, would require too much livestock. As I have attempted to demonstrate, this form of attack misuses its own device of argument, ignores the highly relevant subject of where livestock production is now and could be located, and fails to acknowledge that a surprisingly modest amount of livestock can be organized to create a threshold of viable organic and conservation methods.

Livestock and the Environment

To close this general discussion of the place of livestock in sustainable agriculture, I turn briefly to another broad set of criticisms of livestock production, those based on environmental, health, or ethical considerations. The critics generally attack the production and consumption of meat on one or more of a variety of grounds, including animal suffering, personal health, pollution, and inefficient conversion of plant to animal protein. Typical is a recently formed group called the Beyond Beef Coalition. In a full-page advertisement in the *New York Times* (April 14, 1992, A9), the group announced a campaign to reduce beef consumption by 50% by the year 2002.[12] However, in small print the sponsors state that they are not opposed to beef production in organic and otherwise sustainable systems. This disclaimer is too important for small print. It deserves amplification.

A detailed discussion of the many issues raised by such critics is beyond the scope of this chapter. However, I will make one suggestion for advancing the debate. It is that critics should recognize the distinct differences between and possibilities of two contrasting modes of cattle production: cattle in concentration and cattle on farms.

Raising cattle in dense concentrations in feedlots contributes to several of the situations which are cited as problems. Animal welfare is compromised by exposure, crowding, disease, and restricted freedom of movement. Furthermore, cattle manure becomes a pollutant because it is uneconomical to transport it to any but the nearest agricultural lands. In addition, feedlot cattle do not convert plant to animal protein nearly as efficiently as they could, since they are prevented from foraging grass and crop residue, two key examples of plant protein not usable by humans.

In contrast, when cattle are widely dispersed on farms in modest numbers, there are no inherent impediments to addressing these criticisms. Regarding animal suffering, cattle on farms can be dispersed with freedom of movement, can be given ample protection from exposure, and can function in situations much as they would in the absence of human intervention. Regarding animal pollution, manure can be managed where cattle are modest in number and dispersed. Regarding conversion of protein, it is useful to remember that cattle are ruminants; they can utilize crude protein not digestible by humans. Thus on farms, they can forage on soil-conserving grasses and consume soil-enhancing legumes.[13] In fact, cattle could be raised entirely on such a diet. The limiting factor is what consumers will accept. And finally, regarding the charge that beef is high in fat and therefore unhealthy, cattle which are fed forages rather than grain will yield much leaner meat.

Obviously, raising cattle on farms rather than in feedlots does not guarantee that they will be better managed. What should interest us is that, on the farm, cattle production can be organized to avoid or minimize many of the problems that critics point to in advocating their antilivestock positions.

Suppose, as I have argued, that a modest livestock component is essential to organic farming, as well as other alternative systems.[14] Suppose, further, that nutritional concerns can be satisfactorily addressed and that livestock production on farms can be organized to avoid the sort of problems just described. If so, then not all livestock systems should be regarded in the same way, and groups such as Beyond Beef should be more careful in specifying their targets. Criticism should be directed at the practice of raising cattle in large feedlot concentrations as well as other types of mismanagement. Those who are committed to environmentalism and to animal welfare should take a positive and supportive interest in well-organized livestock constituents of organic and related farming systems.

A Special Forage Opportunity

To conclude this chapter on livestock in sustainable farming systems, I turn sharply from somewhat abstract argumentation to two specific topics of practical interest to the livestock-raising farmer. I first discuss turnips to illustrate a central point of this chapter. Then I describe a method of incorporating a cow-calf herd into a farming operation so as to preserve maximum acreage in grain crop production and to avoid other problems.

Thus far in this chapter I have referred to the idea of organizing crop and livestock production in a single operation in a way that is mutually reinforcing. This important concept needs a practical concrete illustration. My experience with turnips can make the idea vivid.

In addition to being a highly palatable, nutritious, and high-yielding forage crop, turnips offer a number of other important benefits to the organic farmer. For one thing, they provide the opportunity for double cropping. Turnips are ideally suited as a second crop after summer grains and clover have been harvested and even following some fall-harvested crops. Second, in multiyear dry periods, moisture-depleted fields of clover or other crops harmed by drought can be planted to turnips rather than to row crops. Third, fields destroyed by insects or disasters such as hail can be replanted to turnips as late as mid August. Fourth, the weed-suppressing effect of turnips helps to control weeds in the succeeding year's crop. And finally, rotation from turnips to soybeans and milo has repeatedly augmented yields, although I do not understand why. All that is required to take advantage of these considerable benefits of turnips is to have livestock to utilize the crop.

The feed value of turnips is impressive. The roots contain 14% crude protein, 0.64% calcium, and 0.22% phosphorus; the total of digestible nutrients is 84% (Neumann 1977, 851). The nutrient analysis of the tops is similar. Turnips thrive under a wide range of temperature and light conditions. They grow well from the northern tier of states to Florida, they continue growing even as daylight shortens in the fall, and they resist frost damage to 18–20 degrees Fahrenheit (Zahradnik 1984, 32).

Raising turnips is fairly straightforward. I sometimes plant them in the spring overseeded into oats or wheat. More commonly I plant in mid summer after a wheat or oat crop has been harvested. The ideal planting date in eastern Nebraska is the last week in July. Turnips are also a good crop for set-aside acres, but in recent years the Agricultural Stabilization and Conservation Service has not granted permission to plant them.

The best tillage for turnips is moldboard plowing, but disking is also acceptable. An advantage of disking oat stubble is that the resulting volunteer oats are a superb companion to the turnips. For spring seeding, I use a mixture that includes sweet clover and red clover in equal proportions. Turnip seed is commonly available in 50-pound sacks at approximately $1.25 per pound. My experience has been with the purple top, white globe variety. I have experimented with mixes that also include such things as rape and rutabaga; however, the best results have always occurred when turnip seed is predominant.

Since a seeding rate of more than one pound per acre is not desirable for late summer seeding, a grain drill will not work. Therefore, I use a broadcast seeder. Perhaps the one challenge in raising turnips is keeping the seeding rate low enough. The temptation to harrow after seeding should be resisted because that places the seed too deep. A shower is sufficient for sprouting.

I have planted as many as 110 acres of turnips in one year. The spring-planted turnips are allowed to grow along with the clover until fall. The summer-planted turnips are usually ready for forage just as the fall crops are harvested. I let the cows and yearlings graze freely. They quickly become enthusiastic, so much so that ultimately everything is consumed. At first cattle not familiar with turnips will eat only the tops. Then about a week or so after the tops are gone, they discover the bottoms. This intervening week has needlessly discouraged some producers. The cows continue foraging into the winter, sifting through the snow and eating partially frozen bottoms. Even after thawing, turnips remain palatable; cows devour the mushy, smelly remains.

The highly nutritious and palatable nature of the turnip offers several advantages. It is well suited to both cows and their calves in the fall. Cows have an opportunity to get into good shape for winter without foundering, while calves make impressive weight gains. High palatability assists in managing a problem of fall pasturing on a diversified farm. That is the tendency of cows to overgraze fall-planted wheat and alfalfa before dormancy is completely assured. This problem is greatly reduced by the appeal of turnips to the cattle.

Various problems have been reported with growing turnips or with using them for cattle feed. One is that turnips are very sensitive to herbicides, thus providing another incentive to be free of them. Another is that, in the West, some cows which are turned from sparse grass pasture to lush turnip fields have suffered emphysema, a degenerative brain condition, and a gas-free bloat from overeating (Effertz 1985). A third reported problem is that dairy cattle have developed a thiamine deficiency if the turnip diet is not supplemented with roughage (Effertz 1985). Fourth, there is some talk of the danger of cattle choking on bulbs, though I have not heard of any reported cases. I have experienced none of these problems, perhaps because my cattle range freely over a variety of crop residues. A final concern is that cattle eat turnips so voraciously that most of the protective ground cover can be removed. In some cases it may be necessary to do some supplemental seeding using grains such as oats or rye.

Returning Livestock to Prime Farmland

Next, I will describe a method of integrating livestock into a farming operation which still retains a high proportion of the land in crop production. Imagine a farmer cultivating land that is well suited to cash crops such as corn and soybeans. Suppose further that he or she is attracted to low-chemical, conservation-oriented farming. But the farmer—and perhaps the landowner—is not eager to convert soybean and corn acres to pasture. The dilemma is heightened by the "sodbuster" provisions of the 1985 federal Food Security Act, which sometimes prohibit fields seeded to grass from being returned to row-crop production. I will describe a beef cow-calf operation that is appropriate to such a situation. It is a modified summer dry lot system for a herd of 20 to 35 cows.[15] The spatial dimensions of the farm are illustrated in Figure 4.

The operation is organized on a two-year cycle. Through the winter, the

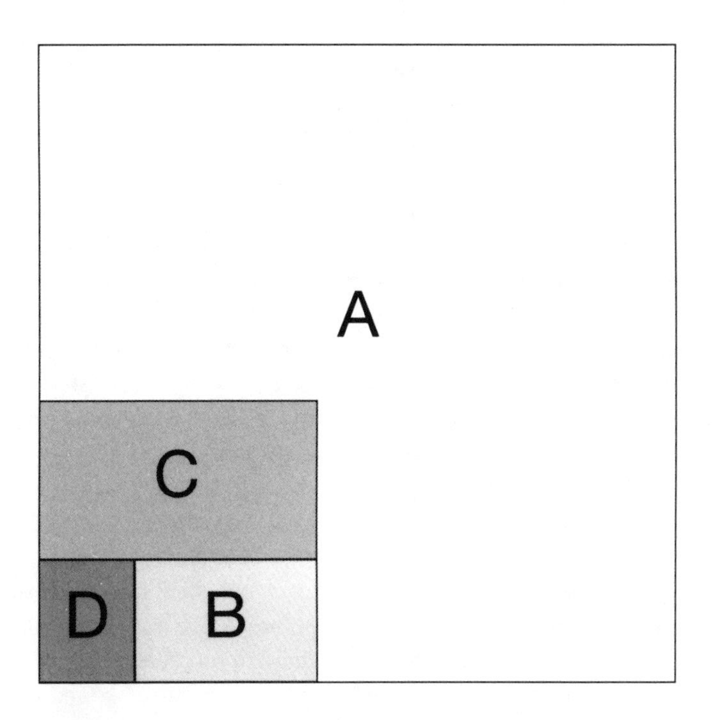

A: A 160-acre tract with permanent fence or provisions for electric fence on all sides.
B: A 3- to 4-acre fenced lot seeded to permanent grass.
C: A fenced field of about 15 to 30 acres which takes in any rough, steep, wooded, or creek land on the farm.
D: A farmyard or simply an area with holding pens, water, and hay storage.

Figure 4. Farm layout for a modified summer dry-lot beef cow-calf operation.

cows pasture the entire farm and any tracts beyond which are available. When the cows are taken off the fields in the spring, they are placed in D and/or B for calving and are kept there until early summer. Tillable land in C is planted to oats. After the oats crop is harvested, the herd is let into C for the remainder of the summer. Feed sources there include oat stubble residue, forage crop that was planted with the oats, and grass from untilled acres. However, some supplemental feeding will most likely be required. After the oats stubble is thoroughly grazed, terraces can be plowed to maintain their structure and the remaining stubble disked. This tillage will kill weeds and start a new growth of volunteer oats for further pasturing and for winter cover. After the fall crops are harvested, the herd returns to all four areas.

During the second year of the cycle, the tillable acres of area C are planted in short- to medium-season soybeans. If planting is delayed until late May or early June as part of the weed control strategy, C will be available until then for calving and for grazing of its grass areas. Through the remainder of the summer, the cows are confined to D or to D and B. Since the soybeans on C are a short-season variety, this field will be one of the first harvested in the fall, allowing it to be opened for grazing early, possibly even in September.

Although there are periods during the two-year cycle when the herd must be confined and fed, this system does have a number of considerable advantages. One is that the farm has almost as many acres in cash crops as before. The 3 to 4 acres in B are committed to permanent grass so that in a wet calving season some grass turf is available. But if field C includes areas that should not or cannot be tilled, these become productive in this system.

A second benefit is that this livestock system makes possible a rotation which contributes to nonherbicide weed control. Oats and soybeans are highly compatible in rotation, and alternating a row crop with a sod crop avoids the weed pressure associated with planting row crops or annual sod crops in successive years. For example, in the oats year, summer tillage after harvest can deal with cocklebur and velvetleaf, two weeds that are promoted by soybean culture. And the soybean year permits spring tillage of pennycress and wild oats (*Avena fatua* L.), weeds which often accompany oats. Furthermore, using this rotation instead of having permanent pasture helps in controlling noxious weeds, such as Russian thistle (*Salsola iberica* Seun. & Pau) and musk thistle (*Carduus nutans* L.), which tend to establish themselves in pasture. Finally, a cropping sequence that includes an annual sod crop in alternate years is a highly soil-conserving practice in the soil-loss formula of the Soil Conservation Service. The rotation, especially if combined with terraces and waterways, will most likely meet Soil Conservation Service requirements for acceptable farming practices on land designated as highly erodible.

A third advantage of this livestock raising system is that there is an extensive period of time in which manure falls directly on a field in crop production rather than on the feedlot or pasture. In the oats year the herd spends the last half of the summer on C, and in the soybean year the herd is on C for most of the spring and from early fall on.

And finally, a fourth advantage is that this system helps the farmer avoid running afoul of the so-called "sodbuster" provisions. These regulations stipulate that, as a condition of eligibility for federal farm programs, certain kinds of vegetative cover must remain intact in certain situations. Sometimes this includes newly planted pastures. Thus, the three or four acres of field B might come under this provision, though on some farms, B can be created on land already in permanent grass. However, because field C remains in annual crop production, it will not come under the sodbuster restriction, thus preserving cropping options.

This modified dry lot management system which I have described could be adapted in various ways. As examples, let me suggest two possible modifications. One would be to fence in two C-sized fields. This would permit both years' utilization to occur in the same year, one in each field. As a result, cows would have grazing areas every month but June of every year.

Another variation would be to plant field C to forage sorghum and use it for summer or fall pasture. This could occur in either year of the rotation. Although there is no cash crop on C in that year, summer feeding requirements are substantially reduced. Operators who do not own the farm could pay cash rent for the acres in C in years when it is planted to forage sorghum. In either fall, the field could be planted to rye, which would increase forage and winter cover and provide an area for spring calving.

If later the farmer decides to make more significant changes in the system, the fences could still be useful. For example, farmers who later decide to have permanent pasture could simply use the fenced areas. Or if it is decided to commit a portion of the farm to the federal Conservation Reserve Program, but to continue raising cattle, field C could be put in the reserve, and a new field C laid out adjacent to it, incorporating one of the fence lines as a side of the new field.

There are many obstacles to returning livestock to farms. The modified summer dry lot management system for beef cows and calves overcomes a number of these difficulties and provides many of the significant benefits associated with raising livestock. Though the details of it may not be appropriate to all situations, it is likely that similar systems could be devised for other livestock, especially ruminants.

Conclusion

Livestock in alternative agriculture is a vast subject. My project has been a modest one, to clarify the importance of that relationship, to touch upon one aspect of livestock and the environment, and to offer some practical suggestions. Implications of this discussion, however, are not so modest. For example, if alternative agriculture is worth pursuing and if I am correct about the importance of livestock to it, then we have ample incentive to create agricultural policy which encourages the reintroduction of livestock to farms.

5

• • •

Comparing Systems

The strengths of organic farming are too often understated and its weaknesses overstated. As a result, it is not taken as seriously as it deserves to be as a viable alternative to the conventional agriculture in place today, the system that is creating such a wide array of problems. I would like to try to combat this weak image by comparing organic and conventional systems from a variety of important perspectives, determining as fully as possible the advantages and disadvantages. Not all the advantages of pesticide-free systems are derived solely from their pesticide-free aspect. Sometimes they emerge from the preconditions of pesticide-free farming, such as crop rotation, diversification, and livestock. These components of systems are thereby defended as well. In my analysis, I associate livestock with organic systems. The rationale is that while livestock are not exclusive to organic farming, their numbers have declined precipitously on conventional farms and they are generally considered indispensable on organic farms.

Convenience

A basic consideration in agriculture is weeds. Today, the conventional wisdom is that for controlling weeds in row crops, chemicals are more convenient than are nonchemical means. I will challenge this assumption by comparing the two approaches within the context of a complete planting season, looking at the advantages and problems associated with each.

One point of comparison is machinery. Nonchemical weed control does not require any of the equipment used for applying herbicides. On the other hand, chemical weed control at least occasionally requires all of the weed suppressing machinery used for nonchemical control. Some combination of the following spraying equipment is needed for chemical weed control: booms on tillage equipment, tanks on one or more tractors, a general sprayer, and liquid handling equipment. In addition, almost all farmers using chemicals will also need a row crop cultivator to deal with failure of weed control. And they will have to make provisions for a rotary hoe, whether through owning, renting, or custom hiring, to cope with crusting and the effects of herbicide failure.

Another preplanting decision will be affected differently depending on whether or not herbicides are used. That is the location of crops on the farm. When a chemical control system has been in place, the options are much more restricted because of herbicide carryover. The problem is choosing a crop for this season that is compatible with last year's herbicide. The farmer is sometimes prevented from planting a preferred crop in a particular field or may see stressed crops when compatibility is misjudged. One example is the incompatibility of oats and triazine-based herbicides. Farmers who do not use chemicals, of course, face no such problems.

The basic purposes of spring tillage are weed control and preparation of the seedbed. However, with chemical weed control, an additional consideration is incorporating any preplant herbicides that are used. With respect to convenience, the problem is that this cannot always be combined with tillage performed for other reasons. Thus, it is sometimes necessary to till a field just to incorporate an herbicide and sometimes to till it again to enhance the incorporation. Nothing more is gained.

Another aspect of spring tillage is the thoroughness and attention to detail required of each system of weed control. For farmers not using chemicals, the final tillage before planting must be thorough both for weed control and for uniform germination. However, the degree of thoroughness required in chemical weed control is probably even greater. The application of the herbicide must be complete and uniform, requiring accurate mixes and precisely calibrated sprayers. Missed spots result in no weed control, and overlaps increase costs and can stress the crop.

Both methods of weed control can cause stress to the crops. In nonchemical control the rotary hoeing(s) can tear out some of the seedlings. In chemical control the potential problem is herbicide stress. However, compensatory measures are possible only in nonchemical systems. The thinning caused by rotary hoeing can be offset by simply increasing the planting rate. However, doing this with chemical weed control results only in more plants being stressed by the herbicide.

The planting season is an extremely busy time for farmers, no matter which approach to weed control is used. However, the situation is complicated and aggravated for those using chemicals. Herbicides will have to be applied at some stage, whether with the tillage unit, during the actual planting, or before emergence. In addition, if the label instructions and Environmental Protection Agency guidelines are being adhered to, the applicator must wear cumbersome protective clothing and take other precautions. By contrast, farmers using nonchemical weed control are free of these concerns at this acutely busy time.

Shortly after the crop is planted, farmers using nonchemical weed control must begin the fine-tuned field work that is crucial to success. However, the advantage of this method is that the procedures can be carried out as needed, thus saving time and money when weed conditions are favorable. For example, when rain is delayed after planting, I have been able to control weeds effectively with one rotary hoeing and one cultivation. In contrast, farmers relying on chemicals will already have committed the cost of the herbicide before they are able to evaluate the postplanting weed conditions. Thus, they do not have the flexibility to take advantage of opportunities for cost savings. Or consider a problem that can develop after planting, that of crusting. For farmers using chemical weed control, rotary hoeing to break the crust is a wholly additional cost. However, for farmers using nonchemical control, the rotary hoeing is a component of the system. Hence, the crust breaking adds no cost.

Another problem that must sometimes be dealt with shortly after planting is an early crop failure resulting from such causes as heavy rain, hail, lost weed control, plant diseases, or insects. Here again, farmers using chemicals will have applied them for naught, whereas farmers not using chemicals will have avoided some weed control costs. A second potential difficulty is herbicide incompatibility between the lost and the replacement crops. Suppose, for example, that a field of sorghum in the Corn Belt is destroyed by hail on about June 15. The obvious replanting option is soybeans. However, there is a good possibility that soybeans will be incompatible with the herbicide selected for the milo. Even if it is decided to replant with the same crop, it is difficult to determine whether enough of the herbicide remains or whether a second application might be too much.

Two other notorious problems are specific to chemical weed control. One is that the herbicide may drift, damaging the farmer's own or a neighbor's crops. A second problem is the increasing resistance of weeds to herbicides, a phenomenon which is affecting more kinds of weeds and expanding to more areas of the United States.

In summary, chemical weed control creates a host of potential problems for raising row crops.[1] The problems peculiar to this system include crop rotation restrictions, greater machinery costs, possible extra preplant tillage, small margin for error in application, herbicide-stressed crops, and application demands at the busy planting time. This is not to say that nonchemical weed control is without significant challenges. These include the need for an appropriate diversified rotation, the need to extend the planting period so as to accommodate the work load of mechanical weed control, and the need to meticulously time and apply the procedures. Furthermore, extremely wet growing seasons increase problems and reduce options. However, it is important to note that these challenges are not unsolvable and that they are generally met by applying new and refined farming procedures and management skills rather than by making additional capital outlays.

It is not clear how to tally which system of weed control "wins" this competition. I have tried to show that chemical weed control, which is portrayed as convenient and easier, actually causes numerous complications, inefficiencies, and unresolvable problems that do not plague chemical-free weed control on row crops.

Soil Conservation

A second basic area of comparison between organic and chemical agriculture is soil conservation. Although farmers who practice either method may care about land stewardship, it does not follow that the two types of farming are equally conducive to good soil conservation. I will attempt to show that organic systems have more potential to conserve soil by identifying the central components of comprehensive soil conservation—the structures and the practices—and then seeing which of the two farming systems is more conducive to each component.[2]

But first, I will be more precise about the evaluative measure "more conducive." We will say that agricultural system A is more conducive to a soil conservation structure or practice than system B if (other things being equal) it meets at least one of the following conditions:

1. A permits the structure or practice, but B does not, or does so only to a lesser extent.
2. In A, the structure or practice is a by-product of completing other objectives, whereas in B it is not.

3. A provides incentive(s) or minimizes disincentive(s) for the structure or practice, whereas B does not.
4. In B there is a conflict in attaining two or more of the structures or practices, whereas in A there is not.

CONSERVATION STRUCTURES

First, let us see how organic and chemical agriculture compare with respect to the following soil conservation structures: water diversion terraces, grassed waterways, and grasslands.

Water diversion terraces, including both depressions and mounded soil, are constructed along contours to slow and channel runoff. Although such structures are equally compatible with the general cropping requirements of both systems of farming, they are more readily maintained within the organic framework. Many forces compromise the structural integrity of terraces, including wind, water, and tillage. The easiest and most economical method of restoring and reinforcing them is to plow the front and back slopes. Within an organic farming system, the diversity of cropping permits this procedure to be done in the spring, in the case of clover, or in the summer after the small grain crop is harvested. In these situations, there is ample time for erosion-fighting vegetation to become established before winter.

However, conventional farming systems do not include legumes or summer-harvested crops. Therefore, the terrace plowing must be completed in the spring or fall. In the spring the soil is seldom dry enough to permit the prior year's corn or soybean fields to be plowed without creating unacceptably cloddy soil. Furthermore, in the spring, conventional farmers are pressed for time to prepare for summer row crops on all their planted acres. Thus, most terrace maintenance plowing is done in the fall. Unfortunately, this eliminates a portion of the erosion-minimizing residue and leaves bare the terrace crests, which take the brunt of the winds. In some cases, the total portion of field without cover may be 40%. For example, fields with significant slope may require terraces every 75 feet or even closer. If, as is typical, each terrace has a 15-foot front slope and a 15-foot back slope, this means that as little as 45 feet of untouched residue remains for every 75 feet of field.

Thus, conventional farming is less conducive to soil conservation because there is a conflict in attaining two or more practices or structures (condition 4)—in this case, the need to maintain terraces and the need to have winter ground cover. This is not the case in organic farming, since terraces can be maintained at other times of the year so that winter ground cover is preserved.

A second crucial soil conservation structure is grassed waterways for routing runoff. They may be present either as companions to terraces or by themselves. They are somewhat compatible with both organic and conventional systems. A major obstacle to the use of waterways is that the acreage devoted to them is not in active crop production. One way to reduce this disincentive is to harvest and sell the forage. However, this is complicated by the uneven quality of the differing types of grasses and by the need for specialized machinery. But on an organic farm, the livestock provide a ready way to utilize the forage, and the necessary equipment is already part of the inventory. A second major obstacle to grassed waterways is that the grass herbicides used on corn and soybeans can severely damage the sod of the waterways. While some farmers do manage to use herbicides and preserve good waterways, the general incompatibility requires continual vigilance.

Thus, once again organic farming is more conducive to a soil conservation structure—in this case by providing an economic incentive for grassed waterways (condition 3) and by creating a more compatible context for them (condition 1).

Similar considerations apply to land not suitable for crop production because of its high susceptibility to erosion. The effective alternative, which almost completely eliminates erosion, is to let the land remain in grass or to return it to grass. Again only organic farming, with its livestock component and the necessary forage equipment, provides the incentive for doing so.

CONSERVATION PRACTICES

Having looked at the degree to which organic and conventional farming are conducive to permanent soil conservation, I now examine their relationship to soil conservation practices within the framework of crop production.

Other things being equal, sod-base crops, such as legumes, wheat, and oats, permit less erosion than do crops planted in rows. On erosion-prone hillsides, an especially effective strategy is strip cropping, that is, alternating strips of sod-base and row crops. The sod-base crop partially arrests the momentum of the water runoff and partially contains the silt runoff from row crops. Here there is a major contrast between organic and conventional systems. The diversified crop rotation of organic farming always includes sod-base crops. And the need to place several different crops in any farm unit almost guarantees that there will be some strip cropping. In comparison, conventional farming has no sod-base crops, thus eliminating the possibility of effective strip cropping (condition 1).

A second important soil conservation practice is contour planting. The strategy is to place crop rows across and around slopes perpendicular to the flow of water, thus retarding its momentum. This conservation practice is equally well adapted to organic and conventional systems.

Avoiding soil compaction is not only a general agronomic goal but also a consideration in soil conservation. Compacted soil is less able to absorb rainfall, thus increasing water runoff and erosion. The difference between the two farming systems in causing soil compaction is minimal, or at least inconclusive, for they each have distinct sources of compaction. One of the main causes of soil compaction is tilling wet soil. In conventional farming, because all crops are planted in the spring, the likelihood is great that all the field work cannot be completed just in the periods when fields are suitably dry. Wet springs and large acreages exacerbate this problem. An organic farm will ordinarily not have this problem; the diverse crop rotation insures that a much smaller percentage of the total acreage will be worked in the spring. Unfortunately organic farming has its own soil compaction problems. Weed control in row crops requires timely field procedures. However, when rains and timeliness conflict, it sometimes is necessary to go after the weeds even in wet soil. Thus, it appears that neither farming system has the advantage in avoiding working wet soil.

There is, however, another contrast. Other methods of coping with soil compaction include deep tilling, planting deep-rooted legumes and permanent grasses, and improving general soil tilth. In conventional farming only deep tilling is available. By definition, legumes and grasses are absent and soil building is limited by the lack of diversified crop rotation and by the lack of nutrient recycling, such as manure spread on fields. The exact advantage of an organic over a conventional system in these respects will require research to determine. Nevertheless, the advantage, whatever its magnitude, illustrates conditions 1 and 2 in favor of the organic system.

Still another important soil conservation practice is conservation tillage, the goal of which is to leave as much residue on the surface as possible to impede erosion. Both organic and conventional systems can include conservation tillage. It is, however, unresolved as to which is more conducive to high residue cover. Modern field cultivators, planters, rotary hoes, and cultivators—implements common to organic farmers' tillage and weed control strategies—can cope with very high quantities of residue. But less certain is whether herbicides can be made to work properly in such conditions. Suppose a farmer wishes to retain very high levels of surface residue but also wants to incorporate an herbicide. With the sprayer booms at the front of the tillage implement, the herbicide will fall mostly on the residue. If an implement such as a field cultivator leaves the maximum residue on the surface, the result, it appears, will be good soil protection, but uncertain incorporation of the herbicide. The residue cannot both remain on the surface, which is the conservation objective, and also be incorporated into the soil, which would appear to be necessary for the soil to make contact with the herbicide.[3]

One of the simplest, least expensive, and most effective soil conservation practices is to leave fields undisturbed after fall harvest and through the winter so that the maximum amount of erosion-impeding residue remains on the surface. In conventional farming there are two strong incentives to do fall tillage. The first is the need to plow in order to maintain terraces. The second is the pressure to do late fall tillage in order to get a head start on the large amount of field preparation needed for the spring planting. By contrast, organic farming has no such incentives to do fall tillage, but has a strong incentive to avoid it—that is, leaving harvested fields undisturbed for winter foraging by livestock (condition 3).

A final soil conservation practice on which conventional and organic farming can be evaluated is the use of cover crops to protect against erosion. In organic farming, the diversified cropping, including legumes and grasses, insures that some of the acres are protected by cover crops. Additionally, an organic farmer can overseed into standing row crops—for example, rye into soybeans—without fear of herbicide damage. This provides not only ground cover for winter but also forage for livestock. However, in conventional farming there are no crops which provide winter cover. The closest candidate is drilled soybeans, which help very little.[4] Secondly, a conventional farmer who wishes to overseed a cover crop into standing row crops must contend with the problem of herbicide carryover. Third, because we are considering conventional farming without livestock, the incentive to use cover crops for forage is absent. And finally, a disincentive for cover crops is the need to till fields and plow terraces in the fall. Thus, organic farming is more conducive to cover crops according to conditions 1 and 3.

A tillage practice which is currently gaining much publicity as a soil conservation technique is no-till cropping, that is, planting directly into the previous year's crop residue without any preparatory tillage. The advantage is that it leaves the maximum amount of undisturbed residue on the surface between crops and during the growing season of the current crop, thus impeding both wind and water erosion. It may also improve soil structure and soil moisture over time. Although some organic farmers who use ridge-till can use no-till up to the point of planting, no-till generally requires the use of herbicides for weed control.

There are a variety of unanswered questions about no-till cropping in conventional systems. One is whether no-till results in adequate placement of fertilizer and humus through the soil profile. Another is whether, in drought conditions, chemicals are less effective and crop yields decline more sharply with no-till. Another is whether no-till crops are more susceptible to diseases. Another is whether no-till creates long-term, broad-range weed and tree control problems. Another is whether no-till is prohibitively expensive. And finally there is the serious question of whether the complete reliance on chemicals creates environmental problems. A recent study in east-central South Dakota examined causes of water contamination from 1981 to 1991. One conclusion is that "vadose zone monitoring measured a faster rate of movement of rainwater from the soil surface to the ground water for the no-till tillage plots than for the moldboard plow plots. The no-till plots delivered higher quantities of water and nitrates to the 6-foot depth even though NO_3-N concentrations were lower under the no-till plots. In one of the three years of study nitrate concentrations were also significantly higher in the ground water under no-till plots" (Kuck and Goodman 1991, 1:4). Because of issues such as these involving chemical no-till and because of the emerging development of no-till organic ridge-till, it is not now possible to decide comparative effectiveness.[5]

Which system, then, is more conducive to soil conservation structures and practices? With the exception of contour planting, which can easily be utilized in both systems, and no-till, about which too many questions remain unresolved, the advantage is clearly with organic agriculture. While many conventional farmers are undoubtedly dedicated to soil conservation, they are employing a system that presents numerous obstacles, large and small, to ambitious soil conservation. It appears that the most effective soil conservation requires something very much like the central components of organic farming: diversified crop rotation, livestock, and nonchemical procedures.

Management and Cash Inputs

A third basic comparison between organic farming and conventional farming is the allocation of capital and management to achieve specific objectives. In general, organic farming replaces some cash inputs of conventional systems with management. While some critics would contend that this is just a fancy way of stating that organic farming trades labor for management, I will argue that this is not so. Organic systems do replace some cash inputs with labor. However, they also replace some cash inputs of conventional systems with pure management. I will try to make this important feature explicit by giving an example from each of the four seasons as well as one general contrast between the two systems. For the comparison, I will use a typical diversified organic farm and a typical conventional, chemical-intensive farm producing corn and soybeans.

The principle at work here is that in order to achieve a particular farming objective, a conventional system will often require direct cash expenditure. However, in an organic system, the same objective is often achieved as an indirect consequence of the method used to accomplish another objective, one also common to both systems.

Soil compaction is a pervasive scourge of contemporary agriculture.[6] In a conventional system, the main way of dealing with it is deep tillage in the spring using implements designed for that purpose: chisel plows or subsoilers. The time, labor, and machinery required constitute a significant expenditure. A typical organic farm uses a very different and indirect strategy. Legumes will be an important part of the crop rotation for a variety of reasons, including fixing nitrogen, providing forage, translocating trace minerals from the subsoil, serving as ground cover, promoting earthworm activity, and improving general soil tilth. However, in addition to meeting these objectives, legumes provide an additional benefit: their deep-penetrating roots and tilth-promoting capacity help to remedy the problem of soil compaction. The other benefits of legumes easily justify the labor and expense of growing them. Therefore, their beneficial effect on soil compaction is a by-product of the general management strategy of including legumes on an organic farm.[7]

Every summer brings different challenges and opportunities for weed control in row crops. A serious defect of conventional chemical weed control is that the expenses and labor have generally already been committed before the actual growing season. The costs are largely fixed. However, a good organic system is able to respond flexibly to varying conditions. The key is staggered planting. The acreage is divided into small units, and as many as several days are allowed to pass between the planting of one unit and another. As a result, the farmer is maximally positioned to cope with a wet season or to take advantage of a dry one. In a wet season, there will be a variety of field situations as a result of the pauses in planting, and there will be, therefore, more time to tend the fields with mechanical weed control. In a dry season, the farmer can simply do no more than is necessary to control the weeds, perhaps only one rotary hoeing or harrowing.

This strategy illustrates the management versus cash contrast in two ways. First, it does not cost more, it requires virtually no more labor, and crop yields are as often as not positively affected by simply making the management decision to carefully organize the planting times. Second, in dry years the organic farmer, who reasonably could have chosen to stagger planting for other reasons, such as spreading work and reducing fall harvest pressure, is positioned to realize significant cost and labor savings simply as a result of management. That contrasts sharply with the structure of chemical weed control.

In the fall, a major focus on most farms is harvesting the grain crops and handling the grain. Good harvest management depends largely on timeliness. On a conventional corn and soybean farm, all crops are harvested in the fall. Therefore, on a medium-sized or large farm, timeliness requires large, modern harvesting machinery; high-capacity grain moving, handling, and conditioning equipment; considerable grain storage capacity; and often extra labor. It is a capital-intensive situation. By contrast, on a typical medium-sized or large organic farm, a substantial portion of the acres are planted to legumes and summer-harvested crops. As a result, there is less time pressure during fall harvest and, thus, less need for expensive harvest equipment.[8] The organic farmer has other sufficient reasons to practice diversified cropping. A by-product is the substantial savings in fall harvesting and grain handling equipment.

A general consideration from fall to spring is preparing the fields for the coming season's crops. Among the concerns for all farms are preventing volunteer plants, managing crop residue, and maintaining and improving soil fertility. On a conventional farm, such problems are likely to be addressed directly by methods requiring cash inputs: herbicides to control the volunteers, shredders on the combine and tillage to reduce crop residue, and purchased fertilizer to increase fertility. Now consider the situation on a typical organic farm on which beef cattle are present primarily to provide financial stability. When the cattle graze on the fields after harvest, they are also addressing each of the problems associated with field preparation. They reduce or eliminate the volunteer problem by eating the grain that would cause it. They ease the problem of crop residue by consuming a portion of it. And they contribute to soil fertility with the manure they drop on the fields. Once again, a by-product of pursuing other objectives is that the organic farmer partially avoids certain cash inputs and labor of a conventional system.

And finally, a central consideration in evaluating competing farming systems is risk management. Conventional corn and soybean agriculture is notoriously vulnerable to uncertainties simply because all crops are row crops which are planted in the spring and harvested in the fall. Farmers who are not willing to accept that much risk purchase crop insurance, which becomes part of the cost of production. In contrast, an organic farm is diversified in a way that minimizes risk. Drought usually threatens crops in summer, a time when corn and soybeans are most vulnerable. However, on a typical organic farm, one or two crops of hay have been harvested by then, wheat and oats are ready for harvest, and the livestock—a significant source of stability—have already pastured on forage. Though an organic system is not immune to drought damage, it will typically fare better than a conventional system. Once again, assuming that there are other sufficient reasons to diversify, the reduction of risk comes as a byproduct of the management decision to do so. This contrasts sharply with the high risk exposure of conventional systems.

These five examples, together spanning the seasons and representing very diverse and basic aspects of farming operations, illustrate how organic farms typically substitute not just labor, but also pure management for cash inputs of conventional agriculture. In each example, the organic system solves problems by using techniques that automatically serve additional purposes.[9] Furthermore, several examples show how the management of the organic farming system may also replace labor in a conventional system.

Productivity

A fourth basic comparison between organic and conventional farming is their overall productivity. I have argued repeatedly that organic agriculture need not concede superior productivity to conventional agriculture, both because alternative systems are now demonstrably productive and because their potential has hardly been tapped. Another dimension to the comparison is how we conceive of productivity. It is possible that sometimes the concept is biased in a way that discriminates against organic agriculture.

Consider the basic planning strategies of the contrasting systems. The conventional system is prepared to take advantage of optimal growing conditions. It does so by the crops it selects, by insurance use of pesticides, by heavy fertilizer application, and by moderate to heavy seeding rates. The idea behind this approach is straightforward: to get the best possible yields in years with excellent growing conditions. It is a strategy to maximize the maximum.

In contrast, the objectives of the organic system are much more complex and diverse. They include the following:

- minimizing costs by growing crops that require fewer purchased inputs and by being ready to avoid certain procedures as opportunities arise
- stabilizing income by selecting crops diversified not only by type, but also by susceptibility to risks, including market fluctuations
- minimizing reliance on the federal farm program or on crop insurance as a financial safety net through the financial stability built into the organic farming system
- producing commodities even in very difficult years by, for example, using decimated crops for livestock forage

The underlying principle is to organize the operation so as to maximize the minimum productivity that can be expected.[10] In years of optimum growing conditions, the organic system will probably not generate maximum revenue. But in years with poor growing conditions, the conventional system will likely fall short in each of the four objectives of organic systems.

In comparing the productivity of organic and conventional agriculture, we must not characterize productivity narrowly. That is exactly what happens when it is measured simply in terms of yields, gross production, or even net profits in the short term. Analyses of financial performance of other businesses examine not just how well they do in optimum circumstances, but how they perform in all conditions which commonly occur. So also is agricultural productivity a function of performance under widely varying conditions. Thus, even in years when an organic system does not compete with its conventional counterpart in net profit, it still has economic strengths of considerable value such as those expressed in the productivity objectives of organic systems listed above.

While it is not possible to make a full and precise appraisal of which system is globally more productive in the long run, some preliminary observations are possible.

- Conventional systems of only corn and soybeans cannot escape the inherent vulnerability of that cropping pattern. On the other hand in optimal years organic systems are now competing precisely on the perceived strengths of conventional systems: high yields and profits. This is occurring through steady improvements in yields and multiple uses and double cropping for the less profitable crops in the rotation. We might say that the maximizing the maximum strategy is intrinsically unable to address the minimum, whereas the maximizing the minimum strategy has at least partial control over the maximum.

• Although the strategy of maximizing the maximum serves individual farmers well in optimal years, it serves society poorly in difficult years. For better or worse, in our society we support struggling farmers. Hence, less stable systems that draw more upon government safety nets in problem years are not desirable from that perspective.[11]

• In regions such as the Great Plains, which frequently have less than optimal growing conditions, the strategy of maximizing the minimum should be more appealing than in those regions blessed with consistently favorable growing conditions. In areas with less favorable growing conditions, the conservatism of the maximizing the minimum strategy could result in more raw production of commodities in the long run than the maximizing the maximum strategy. This would be the case if, other things being equal, the superior productivity in bad years of the maximizing the minimum exceeded the superior productivity of the maximizing the maximum strategy in optimal years. This possibility would be a very useful subject for research.

Experiment and Economic Analysis

One of the most rewarding and intellectually stimulating aspects of developing a pesticide-free farm has been on-farm experiment. Virtually every aspect of the system has been susceptible to experimentation. It is fortunate that it can be a positive experience, for in the 1970s and 1980s it was a necessity. Now, however, the situation is much different. Information about alternative farming is fast accumulating. Converting or beginning farmers may somewhat follow in the path of others if they wish.

The role that experiment has played, however, has a variety of implications. Consider the economic analysis of competing farming systems. The considerable challenge is to structure comparisons with conventional farms to establish fairness.[12] Almost all mature alternative systems have blended the goals of profit and experiment. It may be explicit and formal, such as the inclusion of randomized replicated plots with trained supervision. More often it is informal. It is carried out by alternative farmers who wish to extend (at least their own) understanding of organic and other alternative processes and procedures.

These goals must often be blended or mediated; they are not always entirely compatible. On-farm experiments may compromise part of the net profit of the production associated with the experimental area. Several examples can clarify this uneasy relationship between profit and experiment.

WEED MANAGEMENT

Part of learning the proper use of mechanical weed control implements is discovering parameters: what are the ranges of plant and weed size, soil moisture, implement speed and adjustment, seed population, and allelopathic situations that contribute to effective weed control? Part of establishing a sense of range is to exceed it. Including such treatments can result in reduced crop population or weed control and excessive costs and labor, all of which affect profit. The discussion of bindweed in chapter 3, which describes experiments that suspend crop production, offers another example.

FERTILITY

Part of understanding fertility is to try things. An example is to plant a heavy consumer of nitrogen, such as corn, in a field with alternative sources of fertility. Another example is to see whether clover permits not just one but two successful wheat crops; soil tests alone cannot answer the question because legumes confer benefits not measurable by those tests (Lamond et al. 1988). Experiments such as these can easily result in radically reduced yields.

COVER CROPS

There are perils associated with winter cover crop experiments. They may jeopardize moisture or fertility requirements of the succeeding crop. Pasturing may cause surface compaction. The cover crop may also be inordinately difficult to destroy. Yet it is hard to imagine any alternative to supplementing literature reports and seed company information by simply trying various cover crop ideas.[13]

EXPENSES

Based on soil tests, organic farmers sometimes purchase rock phosphate. Phosphorus management is not thoroughly understood, however. There is a possibility that organic farming more efficiently utilizes or conserves available phosphorus resources (Lengnick and King 1986, 111–12; Cacek and Langer 1986, 27). Even if purchased rock phosphate proves to have been unnecessary, it is counted as an expense.

A problem arises as we compare economic performance on actual conventional and alternative farms.[14] If this project does not acknowledge the myriad ways in which trying things has prevented optimal profit—the experiment factor—it becomes skewed in favor of conventional farms. The above examples suggest that experiments are vital to conversion and to improving mature alternative systems.

Economic analysis of competing systems, therefore, must develop formulas for factoring the impact of experiment. Sometimes that will take on some complexity. For example, suppose that a conversion-minded farmer first proceeds with little or no livestock and then reverses course. Such a decision can have a negative economic impact upon a farm for many years. Economic analysis that does not recognize such aspects will not capture the full economic potential of the system. The methodological issues deserve more work.

What results is a complication in doing economic analysis of actual farms. Failure to acknowledge the experiment factor, however, places alternative systems in double jeopardy. The first is having to bear the costs of self-taught farming in a milieu unfavorable to alternative farming. The second is studies which construe these costs as an inherent defect of that way of farming.

Managing Work Flow

What about overall convenience of the two systems of farming? What is at stake here may be more than mere convenience. A sense of a well-organized system may in some cases make the difference in a farmer's quest to change. It is highly important to reduce the incentives to revert to methods of conventional farming. Suppose it is time to select and use one's method of early weed control. If the rotary hoe is in need of repair and is not the correct width for the row spacing, the farmer is under pressure to return to herbicides. Much can be done to reduce such temptations.

The first step is to appreciate major contrasts in work loads between conventional and alternative systems. That allows a farmer to plan to take advantage of opportunities and minimize drawbacks of a pesticide-free system. Consider the sketch in table 4 of the activities typically occurring through a calendar year on each type of farm.[15]

What is noteworthy is that the work load on a pesticide-free farm is fairly evenly distributed throughout the whole year, whereas on a conventional farm the work is heavily concentrated in fall and spring. The situation has advantages and disadvantages for each system. The advantages of the organic farm work load are several. First, its greater evenness makes more efficient use of the farmer's own labor, makes it easier to integrate family members into the opera-

Table 4. Typical farm activities

Pesticide-free farm	Month	Conventional farm
livestock, discretionary	JANUARY	discretionary
livestock	FEBRUARY	discretionary
livestock, manure hauling, oats	MARCH	early field work
livestock, seeding legumes, manure hauling, spring tillage	APRIL	spring tillage, fertilizer and chemical application, planting
livestock, pasture weeds, manure hauling, tillage, planting	MAY	tillage, fertilizer and chemical application, planting
livestock, planting, mechanical weed control, hay, pasture weeds	JUNE	some cultivating (depending on herbicide weed control system)
livestock, harvest (wheat, oats, clover seed, etc.), hay, straw, manure hauling, cultivating, summer tillage, planting turnips or other forage	JULY	discretionary, weeds
livestock, summer tillage, hay, weed cutting, manure hauling	AUGUST	discretionary, weeds
livestock, wheat planting, harvest	SEPTEMBER	harvest
livestock, harvest	OCTOBER	harvest
livestock, fencing, manure hauling	NOVEMBER	harvest, fall tillage and fertilization
livestock, discretionary	DECEMBER	tillage and fertilizing, discretionary

tion, and makes it more possible to keep full-time hired help busy. Second, because of the smaller proportion of row crops, spring is much less busy. This frees the necessary time for tending to spring calving and for seeding clover and alfalfa. And it also substantially reduces the pressure to till wet fields. Third, fall is also much less busy, both because of less row crop acreage to harvest and

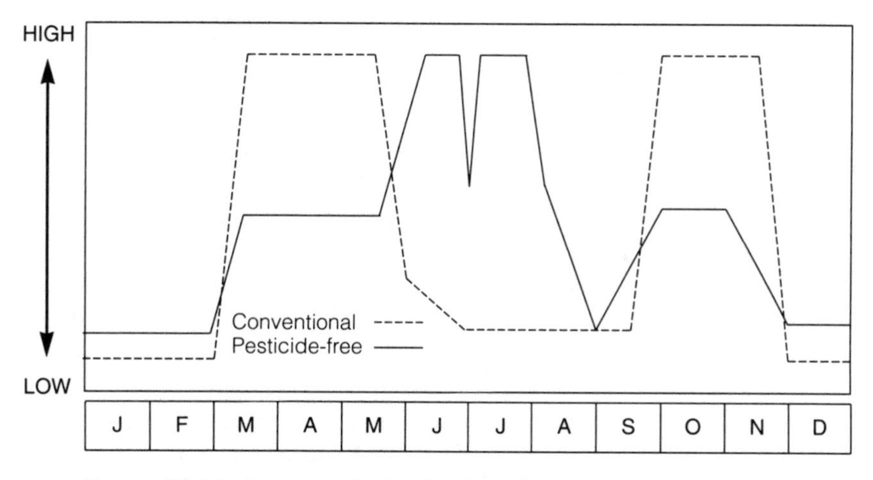

Figure 5. Work load in conventional and pesticide-free systems.

because, ideally, there is no tillage to do. This makes it possible to harvest the row crops with a smaller combine—a considerable savings—and it allows time to do other necessary outdoor work, such as fencing.

I have represented the two work loads on a graph in order to highlight the contrast (figure 5).

Admittedly, there are also challenges posed by the work load distribution of the organic system. For one thing, the work load in early June is heavy because of the need to do planting, haying and mechanical weed control more or less at the same time. Similarly, in early July (though the dates may vary with region) the pressure is strong from summer harvest, haying, straw baling, summer tillage, and late mechanical weed control. And continuing through these times and consistently though the year are the steady demands of livestock raising. The key to system convenience is preparing for these challenges. Fortunately, much can be done.

Early June work-load management. Try to envision the situation. It is June 1. In addition to chores and chopping musk thistles, an organic farmer has many acres of alfalfa hay to put up. There is much planting to do, with its attendant mechanical weed control. The ordinary struggle to find competent help to operate some of the sophisticated machinery is compounded by the additional requirements of pesticide-free farming. What can be done?

Of central importance is having several tractors. A tractor may be needed for haying, for livestock chores, for tillage, for planting, and for weed control. It is often important to have more than one operation done at the same time. Having fewer tractors also increases the chances of a mismatch between tractor size and implement, a situation which can increase ground compaction and fuel consumption. It is especially convenient to leave rotary hoes and planters hooked up to tractors throughout their period of use.[16] I am not proposing that an operation have several new or even late-model tractors. Older 90–120 horsepower tractors, such as the John Deere 4020 and the IHC 06–86, are ideal.

In the note to beginning farmers in chapter 2, I mention machinery which can be de-emphasized. Here, however, is a time to stress the importance of certain pieces of machinery—specifically, a planter, field cultivator, and rotary hoe. I discussed the importance of a planter above in connection with weed control and fast emergence. An additional reason for a reliable large planter is timeliness. One might be justified in having a planter as good as those of conventional farmers who have many more acres of row crops. A new one would not be as expensive, for there will be no need for chemical application attachments.

I discussed above careful preplant tillage. Because of its importance the farmer may wish to personally do much of it. In that case, field cultivator size becomes pertinent. The largest implement affordable and pullable is warranted.

Rotary hoeing and the importance of timeliness have been extensively described. Perhaps this discussion further clarifies the acute importance of timeliness at this particular time of year.

If a farmer is short on help I advise hiring someone to cut the hay and/or put it up. This might be especially useful at the early stages of conversion, as it frees a farmer to concentrate upon the new weed-control techniques. It also may be a way to defer the purchase of certain types of forage equipment. If it is simply a matter of deciding what to hire done, then I suggest the forage work. Most farmers will want to do much of their own nonchemical weed control. Discretionary work, such as hauling grain or working on the combine, can be done at times other than this busy period.

Early July work-load management. Again let us set up the challenge by example. It is July 4 (or whatever date places a farmer in the midst of summer harvest in his or her area). Harvest has just begun. Straw is ready to bale and more wheat is ready for harvest. The oats are almost ready. There are several fields of soybeans at exactly the correct stage for cultivation. Second cutting hay is ready. The farmer will want to till small grain stubble as soon as possible because it has been dry and soil is getting hard. If help is scarce, there is no time to waste.

The above point about multiple tractors applies here as well. On this particular day, a tractor may be required to haul grain, cultivate, cut and/or put up hay, and possibly even pull a tillage implement. Whether a farmer is alone or with help, several tractors are vital.

As one attempts to finish cultivating there will be another benefit from staggered planting. The optimum cultivating stage does not come to all soybean fields at once, so that one can, for example, cultivate in the morning and continue harvest in the afternoon.

If help is limited, consider having someone else operate the combine. That frees the farm operator to organize other projects. In general, this may be the single most desirable time of the year to have help. Part of the explanation, in addition to the extreme workload, is that so many of the projects at that stage of the season can be done by others. Examples are putting up straw or primary tillage of straw fields.

Livestock. The third challenge is livestock. This enterprise undeniably requires year-round work, dedication, and interest, regardless of how well a farmer is organized. Nothing I propose mitigates that. Livestock care is easier on some farms than others, however. There are several reasons why.

Fence building and maintenance is often an unpopular activity. Possible alternatives include using electric fences, hiring help to assist with the fencing, and in some communities, contracting with fence building companies.

Handling facilities are essential for safe, efficient care, for sorting, and for loading. Farmers with the necessary skills can build their own using plans and guides such as the Midwest Plan Service's *Beef Housing and Equipment Handbook* (1987). Or a good alternative is to purchase ready-built handling units, many of which are advertised in farm publications. Feeding areas can include creep- and self-feeders. The hay feeders can be flexible to accommodate different types of hay packages and of large enough capacity to hold several tons at once.

Many farmers have an aversion to manure handling. A way to partially compensate is to be well organized. With several tractors on the farm, it is likely that the manure loader can remain on one of them for extended periods, which will make timely manure handling easier. Furthermore, having a large, good manure spreader hooked to another tractor increases efficiency. And if still another tractor is available to disk in the manure shortly after application, the nitrogen will be retained much better. Renting a skid loader for cleaning confined areas also speeds the process.

Miscellaneous organizational ideas. The keys to making pesticide-free systems more convenient—being organized for livestock and for the the busiest periods—have been discussed. I conclude with brief suggestions for other times of the year.

The complexity of the farming system creates a surprising demand for wagons. The farm will raise a variety of crops, and there will be many livestock feeding situations. Everything from large gravity wagons to barge wagons to old flare-box wagons becomes useful. The farm will greatly benefit from having one gravity wagon with a side-mounted, swing-away auger powered by the tractor hydraulics. It can be used to fill planters and grain drills with such seed as oats, wheat, soybeans, and rye. It can also be used to fill creep feeders and to unload or transfer grain without setting up an independent auger. They are not expensive to buy. Or the parts can be ordered and retrofitted to any gravity wagon.

Pesticide-free farming involves continual seeding of clovers, alfalfa, and grasses. Drag harrows and rolling packers are vital to good seeding. They are regularly available at farm sales. Farmers who have several locations might consider owning more than one of each, for they are often more work to transport than to use.

If a farmer is short of help, I suggest hiring assistance for the forage work. Especially at early stages of conversion, hiring help for forage harvest frees the farmer to concentrate on the new weed management techniques. Another suggestion is that farms with rotations heavy on legumes, grasses, and straw should have a small square baler. And to go along with it, there should be several good hayracks. Sometimes a field of forage can be loaded on racks and then unloaded later when it is more convenient or when help is available. This is especially useful for dealing with straw during summer harvest.

The operator of a pesticide-free farm will be surprised at how easy the pace is in spring in contrast to the pressures on neighbors who farm conventionally. To spread the work load out even more, try to shift tasks to that season. An example is alfalfa work. Seed and destroy alfalfa then rather than in late summer. There are other advantages. Spring seeding with an oats nurse crop better prepares the field for the scuffing of winter grazing. Alfalfa destruction delayed until the next spring permits one more winter of grazing and soil cover.

Conclusion

I hope that this chapter helps to convey some sense of the complexity, subtlety, and wide advantages of organic and other alternative systems. These systems have vast potential to solve major agricultural problems. As apparent as these aspects are to practitioners and researchers, it has not been easy to tell this story. Part of the problem is that proponents are increasingly diverted from this message. Instead, they often find themselves on the defensive. This odd situation is the subject of the next chapter.

6

• • •

The Assault on Alternative Agriculture

It wasn't too long ago that a typical advertisement for farm chemicals would feature a macho farmer eager to smash yet another weed of some sort. How things have changed! Now the pages and screens are filled with trees, birds, deer, wildflowers, young children, soft light and music, and talk of the environment. The macho farmers have been replaced with thoughtful farmers who describe their deep regard for soil and water. One spring 1992 advertisement even claimed that using a particular brand of herbicide is the environmentally responsible thing to do.

An explanation of all this is possible.There is a developing public relations battle over what is appropriate in agriculture. Many segments of agribusiness and allied interests are organizing to fend off the challenge of alternative agriculture. How successful they are will have much to do with the kind of agriculture we have in the future.

Several assertions by opponents to alternative agriculture are at the center of most debates on the subject. The sophistry in these assertions may or may not be apparent. It is clear that they have impact upon the debates. One indication is that conventional farmers often employ one or more of these to justify controversial practices. These assertions are the following:

- Abrupt cessation of chemical use would cause calamity.
- The key to pesticide safety is following label directions.
- Farmers have a moral obligation to feed the world.
- The world is filled with risks.
- Alternative farming requires too much livestock.

These claims are at the heart of the debate. An uncritical acceptance of them might lead one to wonder why anyone would ever want to farm without pesticides. Taken together they place at least some versions of alternative agriculture on the defensive, implying that it is irrational or even irresponsible to try to farm without pesticides.

If, as I intend to show, the arguments behind these assertions are not very good, then why are they effective? I think that there is a smaller and a larger reason. There is at least something intuitive about each of the assertions. There is a connection between morality and feeding hungry people. The world is filled with risks. An *abrupt* cessation of chemical use does give us pause. And even the label directions assertion connects with the commonly held view that if a pesticide were not safe, it would not be on the market. So at least three of these assertions have some surface plausibility. What must be explored is what follows from them. The label directions claim must also receive a much closer look. (The assertion that organic agriculture requires too much livestock has already been addressed in chapter 4.)

The larger reason, I suspect, is that these assertions sometimes indirectly express widespread doubts about or discomfort with alternative agriculture. The most effective response is to simply show how alternative agriculture can work. Still, these assertions must also be considered on their merits.

Calamity

The first argument predicts calamitous results if farmers stop using chemicals. Studies concluding so are regularly reported in the media.[1] For example, here is how the *Omaha World-Herald* summarized a recent study, citing one of its participants, Thomas Foster, an agricultural economist for the Tennessee Valley Authority and the National Fertilizer Development Center: "Taking chemicals out of agriculture could mean cutting exports in half, 217,000 fewer jobs, rampant inflation, and a huge increase in food costs. . . . Yields generally fell under all scenarios while costs per unit of crop increased in all cases" (15 Nov. 1990, 19). Even more dire consequences are sometimes foreseen. Former U.S. Secretary of Agriculture John Block is quoted as warning that "the unthinking critics of . . . chemicals need to realize that they may be trying to force a decision. . . . That decision is, what half of the world eats, which starves?" (*Nebraska Ag Report,* Oct. 1990, 1).

The disastrous results projected in the extreme version sound rather plausible. Suppose that farmers are suddenly asked to stop using the methods they know and are equipped for and told to replace them with methods that they do not understand, have no experience with, and are not equipped for. Chaos might indeed follow. Such would also no doubt be true if a similar, sudden shift were mandated in other sectors of society, such as transportation, medical care, education, or manufacturing.[2]

The problem with the extreme version is that hardly anyone has ever seriously proposed such a thing. Why should they? There is nothing in the rationale for or methods of alternative agriculture that calls for abrupt cessation. Quite to the contrary. Conversion from chemicals is a slow, methodical, step-at-a-time procedure (see Kirschenmann 1988). Some of the features, such as legumes or livestock, can easily be blended into a conventional system, and many farming operations are currently a combination of conventional and alternative methods. Thus, the abrupt cessation calamity is a preposterous straw man argument.

A less extreme version of this general argument is that alternative agriculture would deny society the productivity and other benefits of conventional agriculture. An excellent example is the 1990 study *Economic Impacts of Reduced Chemical Use* by a team of researchers led by Ronald D. Knutson.[3] It was sponsored by a consortium that included the American Farm Bureau Federation, the American Soybean Association, Riceland Foods, DowElanco, Monsanto, ConAgra, R. J. Reynolds, and IMC Fertilizer. The study concludes that reducing chemical use in agriculture will cause yields to decline, food costs to rise, exports to fall, and soil erosion to increase (Knutson et al. 1990a).

In projecting the productivity of competing agricultural systems, it is important to consider whether each system has received approximately equal support for developing its potential. Suppose, for example, that one system had received crucial support while the other had been denied it. The result would be that appraisals of the performance—actual or projected—of the systems would be influenced by that disparity, with the ultimate potential of the less favored system substantially understated.

This is exactly what has happened. The 1989 study *Alternative Agriculture* sponsored by the National Research Council examined the context in which alternative agriculture has developed. Among its conclusions are the following:

> A wide range of federal policies, including commodity programs, trade policy, research and extension programs, food grading and cosmetic standards, pesticide regulation, water quality and supply policies, and tax policy, significantly influence farmers' choices of agricultural practices. As a whole federal policies work against environmentally benign prac-

> tices and the adoption of alternative agricultural systems. . . . A systems approach to research is essential to the progress of alternative agriculture. . . . Little recent research, however, has been directed to alternative agriculture, such as the relationship among crop rotations, tillage methods, pest control, and nutrient cycling. Farmers must understand these interactions as they move toward alternative systems. As a result, the scientific knowledge, technology, and management skills necessary for widespread adoption of alternative agriculture are not widely available or well defined. (6)

In fact, the importance of research to developing an agricultural system is acknowledged by the Knutson study: "The interaction of public agricultural research and extension with private sector research and development has been a primary factor in the increased productivity of American agriculture" (7).

Thus, it seems clear that key support for alternative agriculture has been lacking and that significant obstacles to it exist. Projections based on models distorted in favor of conventional agriculture and distorted against alternative agriculture are of little value. Try to imagine what alternative agriculture would be like in the 1990s and what confidence we could have in its future if it had received even a fraction of the institutional support that has been bestowed on its conventional counterpart.

THE KNUTSON STUDY

I argued above that by itself the recent history of agriculture in the United States is apt to bias current attempts to compare the potential of competing systems in favor of conventional systems. It may still be useful to look at one of these studies; there can be considerably more problems than the one asserted above. I will focus on the Knutson study to raise several methodological questions. Studies that do not extrapolate from actual systems must proceed with their own sense of what alternative agriculture is. The Knutson study is an example. I will, therefore, be keenly interested in how alternative agriculture is conceived in this study.[4]

In appraising a project such as the Knutson study, I am guided by these central questions:

- Does the study disclose the assumptions which underlie the comparisons?
- Does the study display an adequate understanding of alternative agriculture?
- Are well-managed conventional systems compared with well-managed alternative systems?

- Does the study draw upon key information about systems?
- Does the study avoid blaming alternative systems for problems arising from the infrastructure of conventional agriculture?

Assumptions. There is something very curious about the Knutson report. Much of its 72 pages consists of calculations presenting conclusions. The statistical appendices alone extend from pages 51 to 72. The study is also quite explicit in laying out and explaining its assumptions about policy (10–11). But on the critical matter of its assumptions about alternative agriculture—crops, rotations, livestock, weed and insect control, fertility, nutrient recycling, management—the report simply says, "In formulating the yield impacts, the lead crop scientists were asked to consider the potential for changes in cultural practices such as crop rotation, green manure, increased mechanical cultivation and/or hand labor" (9). That is all that is revealed.[5] All of the complex components of an alternative system are referred to simply by that single sentence. How can one begin to understand, let alone appraise, such a project with no more indication than this of what is assumed about alternative agriculture? The following are some issues for fixing assumptions about alternative systems, issues which the 140 crop scientists may or may not have been asked to address in their computations.

A key component of the Knutson report is its projections of serious declines in crop production without chemicals. One set of estimates, for example, is based on removal of synthetic fertilizers. Yet since the report is virtually silent on what assumptions were used, these projections could well be simply the crops and methods of conventional agriculture minus the fertility provided by the synthetic fertilizers.[6] Such a characterization of alternative agriculture is especially false. Who would dispute, for example, that yields from continuous corn suddenly deprived of fertility inputs would drop dramatically? This is about as far removed as possible from the ways in which fertility is managed on alternative farms. Sources of fertility include livestock manure, green manure, legumes, the positive benefits of crop rotation and diversification, sea products, and mined minerals. For example, a well-managed 320-acre organic farm with 60 cows and 50 acres of alfalfa would have much more obvious sources of nitrogen fertility than the same farm without livestock and alfalfa. Projections about fertility will vary greatly, depending on assumptions about livestock and legumes.

What assumptions did the participants in the Knutson study make? What assumptions would be appropriate? Clearly not just anything can be assumed that would favor the alternative model. However, I cannot think of any rationale for completely excluding livestock. Similarly, the study is silent on the possibility of using limestone and mined minerals such as rock phosphate. Since to my knowledge no theory of agriculture opposes these amendments and since they are commonly used in alternative systems, there was no reason to assume deficits.

As with fertility, weed control in nonchemical agriculture is not simply the absence of herbicides, but rather a complex array of alternative measures. But what assumptions about weed control is the Knutson study making when it predicts a 37% decline in soybean yields without chemicals (5), 35% alone due to the absence of herbicides (14). Was this simply the crops and methods of conventional agriculture minus herbicides? Or was this the outcome of following careful and extensive procedures such as I described in chapters 2 and 3? The study does say, in its discussion of sorghum projections, that "weeds were controlled by a crop rotation pattern which included sorghum and alfalfa" (17). That is a start, but only a start.

The Knutson study makes the following general projection about farm income: "Crop producers would experience gains in income, but livestock producer income would drop by an amount that would nearly offset the gain to crop producers" (1). Yet the separation of livestock production from crop production is a feature of conventional agriculture. Almost all alternative systems combine the two for reasons that I have explained in this book. Specifically with respect to income, producing both crops and livestock provides the flexibility to manage resources so as to take advantage of price fluctuations.

The Knutson study further states that "if the crop and livestock enterprises are on the same farm, one side of the farm benefits at the expense of the other" (41). Such a prediction reveals an amazing lack of understanding of the possibilities within alternative systems. Consider, for example, hay and cattle. When the price of hay is low, there is incentive to market more through livestock by, say, retaining more replacement animals or by retaining feeders to finish. Conversely, when the price of hay is high, more of it can be sold directly by culling breeding stock or by selling rather than finishing feeders.

Thus, the Knutson study arrives at its conclusion about profitability in an unacceptable way, that is, by imposing one part of the infrastructure of conventional agriculture upon alternative agriculture—the separation of livestock from crop production—and then proposes that alternative agriculture take responsibility for the unsavory consequences. Those consequences are economic problems for farms when livestock and crop production are separated. This is not valid analysis.

An additional consideration in projecting income is the degree to which the study credits alternative systems with the capacity to replace cash inputs with pure management as discussed in chapter 5. Only if the crop scientists in the study factored this into the assumptions about comparative expenses will projections about income be useful. Considerable understanding of alternative systems is required in order to avoid understating the possibilities for reducing costs.

The idea of a system. Still another weakness of the Knutson study is its failure to appreciate and take into account the importance of system to alternative agriculture. This refers to the manner in which the various components come together in mutually reinforcing ways.[7] Persons appraising alternative agriculture or making projections about it who do not understand or who ignore the idea of a system will produce analyses which are little more than parodies. Let me illustrate this with oats, an indispensable part of my rotation. An analysis which ignores system context would simply calculate the revenues from oats by multiplying the average yield by an average price per bushel. Using a 1992 price of about $1 per bushel results in a rather paltry revenue projection, even if a value for the straw is factored in. In contrast, an analysis which takes into account all the contributions that oats makes to a system such as mine would assign economic value to the following:

- grain, either fed to livestock or sold as seed
- straw, either used as bedding or fed as a supplement to alfalfa hay, especially to dry cows
- double cropping, either followed by turnips or used as a companion crop to establish alfalfa or clover
- weed control, both by allowing weed control procedures after harvest and by its allelopathic effect in the following year
- soil conservation, both as a close-seeded, sod-base crop and by allowing terrace reinforcing tillage at the proper time
- risk management, both through its drought resistance and by spreading the work load

Actual alternative farms. Still another weakness of the Knutson study is its apparent failure to examine actual alternative farms as sources of information. Alternative farms are not merely a theoretical construct; they actually exist. If projections are to be made about such farms, would it not be appropriate to know what is happening on them? I am aware that stating just how such farms relate to the issues of the Knutson report is complicated. For one thing, any benefits that accrue to such farms that would not be available to all farms (such as special markets) would have to be factored in accordingly. Still, the special complexity of these systems would seem to require that analysts use actual instances for information rather than simply making their own assumptions and hypotheses.[8]

Optimal management. The final weakness of the Knutson study that I will consider is its failure to use optimal management for alternative agriculture in formulating its projections. The report claims that "in providing these estimates, the crop scientists included adjustments in cultural practices that were considered to be consistent with optimal management under conditions of restricted chemical use" (43). However, there are a number of reasons to question whether adjustments for optimal management were actually made. In addition to problems of this sort already cited, consider this glaring instance involving projections for corn acreage: "National corn acreage would increase nearly 7 percent from 69.3 million acres to 74.4 million acres under the no chemical scenario. . . . Corn production would increase the most in the Corn Belt, rising from 35.6 million to 38.6 million acres under the no chemical option" (34). Presumably this is to adjust for export requirements, given their projections of reduced yield.

It is seldom disputed that the single most indispensable component of alternative systems is diversified crop rotation. In the Corn Belt this almost always means diverting corn acres to other crops. For one thing, so many of the tillable acres in the Corn Belt are already planted to corn (hence its name) that it would be difficult to attain diversified crop rotation without subtracting from the dominant crop. Furthermore, corn is an obvious candidate for acreage reduction because of its pesticide requirements, especially when grown continuously,[9] its water and nitrogen requirements, and its tendency to promote erosion[10] and to reduce tilth.[11]

The concept of alternative agriculture as nothing more than conventional agriculture minus the chemical inputs that have made it possible is therefore even further exaggerated. The Knutson study supposes cropping patterns even more extreme than in conventional systems by projecting *increases* in corn acreage above that already in conventional systems—increases in the very crop that most necessitates chemical inputs. By projecting an increase in corn acreage under the no-chemicals scenario, the Knutson study reveals, as effectively as I can imagine, its lack of understanding of the fundamentals of alternative agriculture.

Label Directions

The second assertion used to buttress conventional agriculture by opponents of alternative agriculture is that the key to pesticide safety is following label directions. For example, a 1991 DowElanco advertisement in *Successful Farming* for Sonalan herbicide claims that "when applied properly, Sonalan is not expected to reach concentration levels high enough in water to harm you, fish, or wildlife, including deer, squirrels, quail and other birds" (Jan. 1991). Or consider the pledge offered by FoodWatch for Tomorrow, an organization seeking to change perceptions about food safety. The group is sponsored by the Agricultural Council of America, whose officers and directors include representatives from agribusiness and commodities associations. The pledge is a statement of beliefs and commitments to which the organization and its members express allegiance.

> We believe all people have a right to
> - healthful, abundant food.
> - food that is produced and handled safely.
>
> . . . .
>
> Because of these beliefs I pledge
> - to use products properly.
> - to read and follow all label directions.
>
> (Agricultural Council of America 1990, brochure)

The "because" linkage is obviously intended to imply a causal relationship between following label directions and pesticide safety.

Supporting this view of label directions is another industry doctrine: "The dose makes the poison." It is not the mere presence of a toxic substance that is harmful; the critical factor is the amount that is present. For example, a FoodWatch for Tomorrow brochure puts it this way: "Care must be taken not to equate detection with hazard. Because a substance is there, does not make it harmful. It is the dose that makes the poison. Given the current capability to measure infinitesimal amounts of substances that are deemed 'harmful,' it is entirely possible that the term 'harmful' is no longer an absolute concept" (Agricultural Council of America 1990). Thus, what label directions do, according to this doctrine, is to describe how to avoid the threshold of harm.

Such statements as these imply that all of the many issues of pesticide risk—water contamination, applicator safety, residue in food, health effects—are resolved by simply following label directions. Those directions are being made to bear a very heavy load of responsibility. Of course, hardly anything is safe in the strict sense that there is a complete absence of risk. Rather, as Lowrance (1976, 8) points out, safety is a judgement about the acceptability of risk. The terms need not be clarified further, however. For the considerations to follow should have force even if confronted with the most casual sense of safety or acceptable risk.

THE PRELIMINARY ARGUMENT

One consideration is the process of and the meaning of federal registration of agricultural chemicals. According to a General Accounting Office (GAO) report, Environmental Protection Agency (EPA) registration provides "licenses for specified use of pesticide products. A pesticide product registration sets the terms and conditions of the use of that product, including the directions and precautions for use outlined on the product label. All pesticides must be registered by EPA before they can be sold to the public" (General Accounting Office 1986, 137). Yet it would be surprising indeed if there were no more to be learned about pesticides as they are put into use. As a result of such information, the registration status sometimes changes. In the words of the EPA, "New data on registered products sometimes reveal the existence of a problem or a potential for hazard that was not known at the time of registration" (Environmental Protection Agency 1991a, 6). In February of 1990, the EPA had listed over 70 pesticides that have been suspended, cancelled, or restricted (Environmental Protection Agency 1990).

The chemical parathion provides an example of a recent change in registration status. In the fall of 1990, the EPA announced that most uses of parathion would be cancelled and remaining ones severely restricted. Part of EPA's concern was that "parathion use posed unreasonable risks to applicators and mixer/loaders of the chemical and to farmworkers working on sites where the chemical had been used" (Environmental Protection Agency 1991a, 65061).

We must be clear about the significance of a revised or cancelled registration. It is not the case that the registration was altered because of a change in the active ingredient. Rather, the alteration was made because of additional information about the same active ingredient. Thus, by implication, the earlier label directions did not insure safety. In other words, the EPA registration process itself establishes that simply following label directions is not always the key to pesticide safety. Risk assessment is not always complete at the time of initial registration, the point at which label directions are first formulated.

THE LONG ARGUMENT

In order for label directions to provide well-based assurance about pesticide safety, two conditions would have to be met:

- There must be a database appropriate to pesticide risks.
- Pesticide label directions must be consistently derived from that database.

Let us see what the situation is in actual practice.

Lack of adequate database. The adequacy of the database supporting label directions is called into question by deficiencies in at least these four areas: reregistration, tolerance setting, inert ingredients, and synergy.

In 1972 the EPA was required to reregister all previously registered pesticides based on newly established testing guidelines, ones which could better determine the potential for causing health and environmental problems. However, this testing and reregistration process has not been fully implemented. A GAO report gives the following conclusion about the process:

> While much of the population is exposed daily to pesticides in food and the environment, EPA has limited assurance that human health and the environment are adequately protected from possible unreasonable risks of older pesticides. This is because most pesticides used today were initially registered before contemporary regulatory and scientific requirements were imposed. These older pesticides have not been fully tested to determine, among other things, their potential for causing long-term health effects, such as cancer and reproductive disorders, birth defects, and environmental damage. . . . In the meantime, these products can continue to be marketed. (General Accounting Office 1986, 54)

A tolerance setting is "the maximum limit of pesticide allowed in or on raw agricultural commodities, processed foods, or animal feed" (General Accounting Office 1986, 61). The EPA studies potential health risks of pesticide residue with the goal of setting tolerances at acceptable levels. But there have been at least two problems with this process. One is concern about its scientific basis. For example, in 1979 the EPA's own scientific advisory board pointed out the use of obsolete data to make calculations about dietary exposure (General Accounting Office 1986, 69–70). A more recent study focuses on methodological problems which bias estimates of the nation's dietary intake (General Accounting Office 1991a). The second problem with tolerance setting is that many were established without adequate toxicology and residue data (General Accounting Office 1986, 63). In 1977 the EPA announced that it would reassess the tolerances of some 390 pesticides used with foods and registered prior to 1977. Yet in 1986, only 8 had been reviewed (General Accounting Office 1986, 68). Thus, there is more than adequate reason to be skeptical about the assurance of safety in the EPA tolerance levels. As the GAO concludes:

> Missing and inadequate tests have prevented the Agency from completing many tolerance and tolerance exemption reassessments to date. Because tolerance reassessments are dependent on the data received and reviewed by EPA under the pesticide reregistration program, probably not until the 21st century will the safety of all older tolerances and exemptions have been reassessed according to current scientific standards. Until EPA obtains complete data and reassesses existing tolerances, the potential of many pesticide residues to cause genetic change, birth defects, cancer, and other chronic health effects cannot be fully determined. (General Accounting Office 1986, 70)

A third reason to question the adequacy of the EPA's database for registering farm chemicals is its treatment of inert ingredients. These are solvents, thickeners, propellants, and so forth to make the pesticide more useable. Inerts can include such toxic substances as dioxane, formaldehyde, phenarsazine oxide, and triethylamine. The EPA is aware that inerts can be harmful, but according to the GAO, "EPA has only recently begun to review inert pesticide ingredients, although some inerts were known to be hazardous to humans and insufficient information existed to determine the potential health risks of many others. EPA needs to obtain further data in order to determine potential health risks of inerts about which little is known and to protect the public from potentially hazardous residues in food" (General Accounting Office 1986, 89).

Thus far I have pointed out three of the reasons for concern about the adequacy of the EPA's database for establishing pesticide label directions. Because of acknowledged weaknesses in the assessment process, hundreds of pesticides were required to be retested and reregistered. Yet this process is woefully bogged down. According to the GAO, "EPA is still at a preliminary stage in assessing risks of older pesticides" (General Accounting Office 1989, 20). And

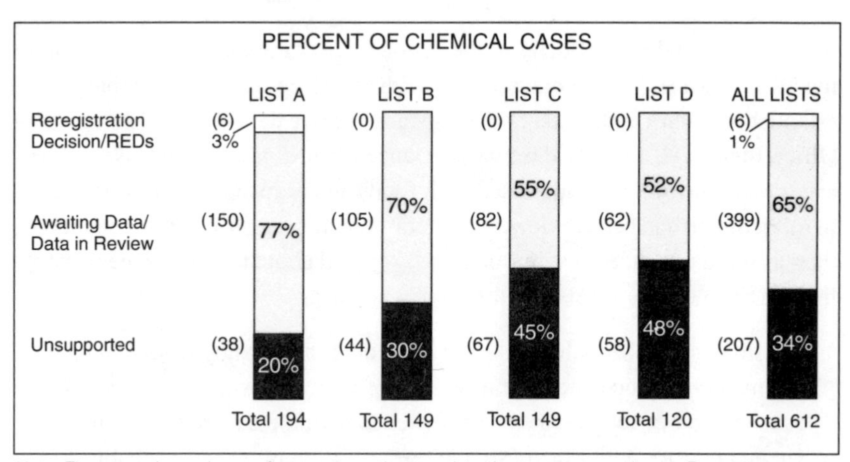

Note: These numbers change frequently as the reregistration process continues. Percentage discrepancies may have resulted from rounding.

Figure 6. Status of chemical reregistration cases, as of third quarter 1991. Redrawn from Environmental Protection Agency 1991c, 1.

even more recent information from the EPA itself reveals that only 1% of all the chemicals awaiting reassessment have been fully reregistered (see figure 6).

One final area in which the EPA's database on pesticide risks is inadequate is in understanding the synergistic interactions of chemicals in mixture, interactions which can result in effects different from the sum of their separate effects. This deficiency is not surprising, for scientific agreement is lacking on when chemical synergy occurs and on the complications of the numerous processes by which it can arise. There has been virtually no animal testing for the effects of chemical mixtures, thus creating severe difficulty for estimating risk. What is clear, however, is that potential for damaging reactions is high. For example, a research group at the Harvard School of Public Health found that ethylpnitropheny benzenethiophosphonate (EPN) and malathion cause a fiftyfold potentiation of acute pesticide toxicity in mammals (Working Group 1986, 12). These researchers conclude that despite the fact that analysis is not yet possible for mixtures that we encounter daily, "we must develop methods to manage the risks associated with exposure to complex mixtures. At least in the cases of gasoline, pesticide residues in food, and drinking water, exposure to mixtures is widespread and frequent. For some compounds synergy is evident" (Working Group 1986, 14). It is clear that at this early stage in our understanding of chemical synergy, the EPA registration process cannot have taken into account this risk in any informed way.

The adequacy of the database is not merely an academic worry, as witness the recent history of ground water contamination by agricultural chemicals. In the 1980s numerous mixtures of pesticides were regularly found in ground water in many states. In a 1988 EPA report, 46 pesticides were identified as leaching into ground water in 26 states as a result of normal agricultural use (Williams et al. 1988). "Normal" presumably refers to generally following label directions.[12] This is just one instance, therefore, where following label directions has *not* prevented readily discernible safety problems.[13]

Label directions not derived from existing data. Thus, the first condition for assurance of pesticide safety—an adequate database—is not satisfied, and on that point alone the claim to safety through label directions is false. The situation, however, is even worse. Consider now the second condition that would have to be met: that label directions must be consistently derived from the database. There are four separate situations in which the EPA is allowed to override the requirement to use an adequate database in registering—and creating label directions for—pesticides.

One procedure is conditional registration. The EPA is permitted by the Federal Insecticide, Fungicide, and Rodenticide Act (FIFRA) to conditionally register pesticides in certain situations, even though some of the required data have not been submitted or processed. This is allowed if the use of the pesticide is in the public interest, if its use doesn't present the risk of unreasonable adverse effects, and if the necessary supporting data are submitted at some future date. This provision, of course, was intended to be used on a limited basis. But the EPA has relied on it heavily. According to a GAO study, "Since 1978 EPA has conditionally registered a large proportion—about 50 percent—of all new pesticide active ingredients even though some of the required health and environmental test data were not submitted to and evaluated by EPA. Since a conditional registration means that a new pesticide is registered for use with less than a full set of required test data, there is some uncertainty with regard to the risks of the pesticide's use" (General Accounting Office 1986, 98–99).

A second situation in which label directions are derived with less than scrupulous attention to the database involves the so-called "Delaney paradox." The EPA acts under the authority of the federal Food, Drug, and Cosmetic Act to set standards for pesticide residues. The requirements of the act are not the same for raw agricultural products, food additives, and animal feed additives. The result is that sometimes cancer-causing residues are allowed on human food. For example, the EPA is permitted to weigh health risks against other benefits in establishing tolerances for carcinogenic pesticides used on raw agricultural products. The EPA has permitted some of these pesticides to be used, thereby exposing people who eat raw products.[14]

A third way in which a pesticide may be used with an insufficient database is through special review. If an active ingredient of a pesticide raises health or environmental concerns, the EPA has the authority to conduct a detailed cost and benefit analysis. The original intention was that the process would be conducted quickly. However, according to the GAO, "Special reviews have taken from 2 to 6 years or longer to complete, and the hearing process which may follow a special review may take up to 2 years or longer" (General Accounting Office 1986, 118). The problem is that while the review and hearing period is dragging on, the pesticide continues to be used, exposing people to potential hazards.

And finally, in some situations the EPA may waive the registration process entirely. A provision of the Insecticide, Fungicide, and Rodenticide Act allows the EPA to grant permission to states and federal agencies to use unregistered pesticides in emergencies. Two serious concerns about this exemption process are the number of emergency exemptions and the number of repeat emergency exemptions. Since 1978, over 4,000 have been granted (General Accounting Office 1991c, 3). In 1990 alone, "EPA granted almost 80 percent of the requests for exemptions for chemicals that had already received exemptions for that particular use for at least three years, and EPA tacitly approved another 18 percent of the repeat requests by not revoking crisis exemptions" (General Accounting Office 1991c, 6). In one instance, the exemption was granted for 12 years in spite of the fact that human health and environmental effects were unknown (General Accounting Office 1991c, 1).

Thus in four separate ways—conditional review, varying legal requirements, special review, and emergency exemption—the EPA's decisions to allow use of chemicals do not derive from a base of information about risk. Hence, this second necessary condition for establishing that label directions assure acceptable risk is not satisfied. Taken together with the failure to meet the first condition—an adequate database of pesticide risks—we can easily see the falsity of the claim that following label directions is the key to pesticide safety.

A FINAL CONSIDERATION

There is one final consideration which further undermines confidence in assurance of safety provided by EPA registration. That is the EPA's use of cost-benefit analysis to weigh known health risks against other considerations. By law, the EPA must assess the benefits of using a pesticide as well as the "costs"—such as health problems—which its use might create (Environmental Protection Agency 1991b, 7). According to Osteen and Szmedra (1989), assessing benefits "is essentially the same as estimating the social efficiency loss, *excluding health and safety effects,* of removing the pesticide from the market and switching to the best alternative controls, if any" (48, emphasis mine). If in this balancing calculus the efficiency benefits outweigh the health and safety costs, the pesticide will be registered or its registration continued. The known health risks are assigned lesser importance. A good example is the special review process conducted for the pesticide alachlor.

In 1985 the EPA conducted a special review of alachlor because studies had revealed statistically significant increases of tumors in rats. Interest in the status of the pesticide was heightened because residues of alachlor were frequently found in water and because it was a very popular herbicide (sold under the brand names *Lasso* and *Dual,* which have similar but not identical formulations), accounting for one-tenth of all pesticides sold in 1986 (Hinkle and Pace 1987, 1). The EPA's scientific advisory panel agreed with its classification as a probable human carcinogen. However, the manufacturer Monsanto urged that the chemical remain unclassified. In the end, the EPA did reregister alachlor. The decision resulted from its evaluation that the benefits outweighed the risks. Specifically, it estimated, according to Hinkle and Pace (1987, 15), that cancelling alachlor would cause efficiency loss of from $235 million to $441 million in the first year of cancellation.[15] The important thing to notice is that the reregistration was made possible not by resolving the safety concerns but by subordinating these risks to other considerations. In such situations it is not at all clear how the specific label directions are then derived. And alachlor is not an isolated instance. According to a GAO study (1986, 109), special review pesticides have been reregistered when no alternative pesticides were available or when alternatives were less effective or economical, even in spite of considerable risks. Examples are lindane and dimethoate, which have been retained for several uses in the perceived absence of alternatives.

Because defenders of conventional chemical agriculture associate the idea of pesticide safety with using chemicals according to label directions, I have devoted considerable detail to examining the database from which these directions are supposedly derived and to the numerous ways in which risks are ignored or overridden in the regulatory process. The very least that can be concluded is that following label directions does not ensure pesticide safety.

Feeding the World

The third argument offered for why organic agriculture is not practical is that farmers have a moral obligation to feed the world. Expanding world population, the argument goes, places even greater demands on American farmers to produce. Conventional agriculture is more productive than alternative systems and therefore it is more appropriate to a hungry world. As the group FoodWatch for Tomorrow put it in a promotional brochure, "Estimated population growth for the next quarter century is so enormous that warning bells should be sounding in every quadrant of the globe. Feeding this population may well be the greatest challenge we have ever faced, one that will test the world's productive and technological capabilities to their very limits. . . . However there is a very serious danger that the world's hungry will not be heard if the environmental and food related agenda is written to meet only the needs of well intentioned but well fed interests" (Agricultural Council of America 1990, brochure). If this argument can be sustained, the result would be a very fortuitous convergence of morality and economic interest. There are, however, at least three problems with it.

The first weakness is the assumption that conventional agriculture is more productive. For reasons that I have already explored in chapter 5 and earlier in this one, proponents of alternative agriculture do not concede inferior productivity, either as organic farming is currently practiced and certainly not as to its potential. As the 1989 National Research Council study points out, "Farmers who adopt alternative farming systems often have productive and profitable operations, even though these farms usually function with relatively little help from commodity income and price support programs or extension" (9). Yet the myth of the superior productivity of conventional systems persists, lending underserved plausibility to arguments such as this one.

The second weakness of the argument that American agriculture is morally obligated to feed a hungry world is that the fundamental cause of hunger is not simply total world food supply, but rather poverty and underdeveloped indigenous agriculture. Recent studies of hunger and Third World development all reach this same conclusion. For example, a 1986 World Bank report titled *Poverty and Hunger: Issues and Options for Food Security in Developing Countries* reports that "projections to 2000 indicate that world food production is likely to keep pace with effective global demand, perhaps even at a lower price (although debt-ridden countries may find that domestic food prices will go up because of scarce foreign exchange). Why, then, do both chronic and transitory food insecurity persist? An analysis of the causes of food insecurity concludes that the main cause is not lack of supply or even high prices. The main cause is the weak purchasing power of some households and nations" (13). And Baum and Lejeune state in *Partners Against Hunger* (1986) that "in low income countries, the poorest 20 percent of the people typically spend 60–70 percent of their income on food, and even then are not able to purchase a sufficient amount" (209). When fundamental problems such as poverty are ignored in favor of simply importing food, the situation is actually worsened. As the authors of *Our Common Future* state: "Thus when countries with untapped agricultural resources provide food by importing more, they are effectively importing unemployment. By the same token, countries that are subsidizing food exports are increasing unemployment in food-importing countries. This marginalizes people, and marginalized people are forced to destroy the resource base to survive" (World Commission on Environment and Development 1987, 129). The solution, according to this report, is to shift food production to food-deficit countries and, within these countries, to resource-poor farmers (129).[16]

In *Hunger and Public Action,* Drèze and Sen (1989) defend this same thesis and illustrate it with a variety of case studies. I will summarize two of these, one showing how a situation of food deficit was affected positively by economic adjustments and one showing how a situation of ample food supplies was transformed into a famine by economic causes.

In the early 1970s Maharashtra was a comparatively well-developed state in western India, but with severe poverty in semiarid rural areas. From 1970 to 1973 a spiral of drought and environmental degradation threatened much of the state. Yet according to Drèze and Sen, the suffering associated with the drought was "remarkably confined" (128). How was famine averted? A major reason "undoubtedly relates to public policies of entitlement protection. The cornerstone of these policies was the generation of employment for cash wages on a large scale, supplemented by 'gratiutous relief' for those unable to work and without able-bodied relatives. . . . In the more severely drought-affected districts, the contribution of wage income from employment on public works to total income was well above 50 percent for most villages" (129). One consequence was that food from other parts of India was imported through private trade. Though food production declined during those years, the "reduction in aggregate consumption compared to ordinary years was distributed remarkably evenly among different socioeconomic groups" (131), reflecting the sustained purchasing power of the impoverished groups.

An example of the opposite situation is provided by the famine in Bangladesh in 1974. Drèze and Sen point out that overall food supplies were highest that year of any year between 1971 and 1976. Furthermore, "one of the famine districts (Dinajpur) had the *highest* availability of food in the entire country, and indeed all four of the famine districts were among the top five in terms of availability per head. Even in terms of change in food availability per head over the preceding year, *all* the famine districts without exception had a substantial increase, and three of the four were among the top six in terms of food availability increase among all the nineteen districts" (28). Among the reasons for this situation were flooding, the resultant loss of employment, extreme fluctuations in food prices, and failure of the government to stabilize prices.

In short, numerous respected studies of world hunger reject the "feed the world" role for conventional agriculture. They conclude that a variety of economic and policy factors, not food supply, are fundamental to understanding famine. This is not to say that food supply might not be a causal factor sometime in the future. However, the world hunger argument has been used to justify current production systems, the basic weakness of which concerns precisely its capacity to maintain productivity into the future. This challenge of sustainability raises grave questions about conventional agriculture as a reliable basis for feeding the world in the future.

The final flaw in this argument is the moralizing itself. Even if conventional agriculture were demonstrably more productive than alternative agriculture, this in itself would not provide superior moral sanction for it. It is true that the following moral proposition has wide acceptance:

- Other things being equal, an agriculture which is productive serves the needs of society.

It is also true that other moral propositions about agriculture are quite plausible:

- Other things being equal, producing food which is free of health-threatening contaminants serves the needs of society.
- Since one generation has approximately as much moral worth as another, an agriculture which is sustainable serves the needs of society.[17]

If each of these three propositions is plausible, they specify an agriculture which is productive, which minimizes pesticide usage, and which conserves resources for future generations. Determining just which agricultural system would achieve the best mix of these objectives is precisely what the current debate is about. Placing an ethical perspective on it recasts the debate, but does not alone resolve it.

Risk

The fourth argument which defenders of conventional agriculture use to buttress their case for chemicals and against alternative systems begins with the truism that the world is filled with risk. Here, for example, is former U.S. Secretary of Agriculture Earl Butz: "Americans live in a risk filled world. . . . It is completely unacceptable to believe that there is no way out of the challenges we face. Unquestionably, there are risks involved, but none so great as the risk that we may quit risking" (Butz 1989, 3). The strategy is to try to provide perspective on risk. In a well-known article, Ames, Magaw, and Gold (1990) caution that "because of the large background of low-level carcinogenic and other hazards, and the high costs of regulation, priority setting is a critical first step. It is important not to divert society's attention away from the few really serious hazards, such as tobacco or saturated fat, . . . by the pursuit of hundreds of minor or non-existent hazards" (86). Another example comes from a special issue of *Farm Chemicals* focusing on tips for industry defenders in "delivering plain talk about food safety." One suggestion is to "explain how everyday life is full of risks: driving a car, taking medicine, walking across the street" (Rachman 1989, 24). Implicit in admonishments such as these is the claim that concerns about agricultural chemicals are out of proportion to actual risk. Supposedly what is needed is greater knowledge of the risks or contrasts with other common risks. According to this view, more than just a little concern is irrational.[18]

To respond to this characterization of risk, I will first discuss in a general way acceptable risk for pesticides in agriculture. Then I will look closely at one central concept in risk assessment: threshold of risk.

CONSIDERATIONS OF RISK IN AGRICULTURE

Rather straightforward guides to or issues in determining acceptable risks in general include the following:[19]

- Are the consequences reversible?
- Is the risk distributed evenly?
- Is the risk assumed voluntarily?
- Are there alternatives to the source of the risk?
- Is the source of the risk susceptible to human misuse?

- Is the risk well understood?

Let us apply each of these tests to pesticide use in agriculture.

A society that decided it no longer wished to accept the risks of air travel or car racing could eliminate them simply by halting the activities. However, with pesticides the situation is very different. Many of the central consequences of pesticide usage are not, in any practical sense, reversible. Consider ground water contamination and the accumulation of pesticide residue in humans. It is well established that pesticide residues are measurable in ground water throughout the United States (Williams et al. 1988, Hallberg 1989, Environmental Protection Agency 1986). The storage of pesticide residues in humans was first monitored early in the pesticide era. In *Silent Spring,* Rachel Carson noted accumulation of residues even in the fat of persons with no known exposure (Carson 1962, 20–23). The important point is that there is no known way to remove pesticide residues from ground water or from humans (Environmental Protection Agency 1986, 4). These and other consequences are not reversible.

A second consideration in risk assessment is how the risks are distributed. Of the variety of pesticide problems, some are rather evenly distributed and others are not. Most persons in our society have some exposure to pesticide residues in food. For other groups, however, such as farmer-applicators, farm workers, and pesticide manufacturing workers, there is additional occupational exposure. The World Health Organization calculates 20,000 pesticide-related deaths per year, many of them as the result of working with pesticides (1990, 86). According to its report, "In some occupational groups exposed to certain pesticides, e.g., farmers or pesticide manufacturers, epidemiological studies have demonstrated an increased risk of lung cancer, lymphopoietic cancers, and possibly other forms of cancer" (88). Thus, it is clear that pesticide usage in agriculture exposes some people to much greater risk than it does others.[20]

A third test is to examine whether the risk is assumed voluntarily. If we were discussing safety standards for ski lifts, we would note that the source of the risk is a nonessential activity and that the risk is assumed voluntarily. However, exposure to pesticide risk is a different matter. Food and water are obviously basic needs, and with few exceptions, we cannot avoid consuming food and water which contain pesticide residues. Furthermore, most citizens are, at best, remote from the political processes that have led to current pesticide policy. Granted, some centralized decision making about pesticide risks is necessary.

But such fundamental policy decisions should also be susceptible to democratic processes.

A fourth question is whether there are alternatives to whatever is creating the risk. Many discussions of risks in nuclear power are brought to a halt by the perception (correct or not) that the main viable alternative is coal-fired plants. Probably one reason we accept high automobile fatalities is the idea that there is no practical alternative. Here two arguments converge. The dogma that there is no alternative to pesticides in agriculture would make us more inclined to accept higher risk. Indeed, an assumption of indispensability of pesticides would transform this entire discussion about acceptable risk. In that sense, the calamity argument—that disaster will befall agriculture in the absence of pesticides—is integral to risk analysis.

However, I not only argued that the calamity argument is false, but the subject of this entire book is to articulate and make plausible an alternative to pesticides in agriculture. This question of plausibility has the most profound consequences for the subject of risk. For if pesticide-free agriculture is plausible, then the subject of risk is transformed into the following question: Why are we needlessly accepting the risks (whatever they are) associated with conventional agriculture?

A fifth assessment to make of a risk is to evaluate the degree to which the risk source is susceptible to human misuse. For example, defenders of nuclear power assert that nuclear plants are managed and operated only by people who have been rigorously screened and trained. Even assuming that this is true, notice how implausible it would be to make the same claim about the use of agricultural chemicals. They are handled and applied not just by professional applicators but by millions of farmers of all ages during busy, stressful times of the year. The pesticides are applied without supervision using a variety of procedures, many of them highly complex, such as chemigation, in which the pesticide is mixed with irrigation water. The sources of misuse include, in the words of the EPA, "improper pesticide disposal, accidental spills, rinsate from equipment maintenance, human error in mixing, intentional excess applications, back siphoning into a well during chemigation, or direct channelling of pesticides through poorly constructed or damaged wells" (Williams et al. 1988, 7). Another possibility is cross-contamination in chemical containers. The EPA attempts to keep track of the frequency of such pesticide accidents with its Pesticide Incident Monitoring System, and it reports the findings.

A sixth and final evaluation of risk is to determine if the benefits and risks are well understood. Assessing benefits is not easy and straightforward. In the case of pesticides, benefits are not simply a quantitative measure of the volume of pests destroyed or economic loss avoided. That total must be mediated by acknowledging the unused alternatives that could have achieved much of pesticide's objectives without the disadvantages. The prime and largely unused alternative, of course, is alternative agriculture.

To consider how well the risks of pesticide usage are understood, let's begin by listing some of the toxicology tests that pesticides must undergo in the EPA registration process:[21]

- acute oral, dermal, and inhalation studies
- two-generation reproduction study
- chronic feeding studies on rodents and nonrodents
- oncogenicity studies on gene mutation, structural chromosomal aberration, and other effects toxic to genetic material
- teratogenicity studies on rats and rabbits
- delayed neuropathy studies on chickens
- plant and animal metabolism studies

One reason that tests such as these are required is that animal and epidemiological studies in other contexts have raised concerns about the active agents in the pesticides. In response to these concerns and to the results of the testing, defenders of pesticide safety point out various problems with the science of identifying and estimating risk. For example, Hattis and Kennedy (1990), while not necessarily defending an industry position, offer these qualifications: "Even if analysts know how much of a substance is in the environment, they can't necessarily predict how much people will actually absorb. People breathe at different rates depending on their level of activity: workers laboring heavily at a construction site, garment workers sitting at sewing machines, and people sleeping in the surrounding community will all receive different doses of an airborne contaminant. . . . Finally, people absorb substances in varying amounts depending on the thickness of their skins and the properties of their nasal mucous, and even on whether they tend to breathe through their noses or their mouths" (158).

Uncertainties about causation are clearly an unfortunate limitation of our understanding. The uncertainties, however, are frequently used to fend off responsibility for alleged consequences of pesticide usage. A logical problem arises. Limitations in our understanding of cause and effect sometimes prevent us not only from proving the harm of pesticides but also from establishing any empirical basis for safety claims. Yet some defenders of pesticides would have it both ways, saying, in essence, that since we can't fully specify the harm they cause, we must therefore be "cautious" about suspecting their safety. But this cannot be. We cannot have both limited understanding of harm and at the same time great understanding of safety.

Thus, by using these six standard guides for assessing acceptable risk, we see that pesticide usage ranks poorly on every count: the consequences are not reversible, the risk is not evenly distributed, it is not assumed voluntarily, the source is susceptible to misuse, the risk is not well understood, and above all the risk is largely avoidable through a readily available and effective alternative: nonchemical agriculture. Viewed from this perspective, we can see that it is not at all irrational to have concern about risks from pesticides, regardless of what other risks we face.

THRESHOLD OF RISK

To conclude this examination of the "world-filled-with-risk" argument in support of chemical agriculture, I will critique a key concept used by the EPA (and others) in regulating risk. That is the concept of "threshold of risk." The assumption is that some risks are consequential only at certain levels. Below that level, risks may be ignored. Here, for example, is how the EPA defines the threshold for its pesticide programs: "The concept of a negligible risk is the attempt to set a standard below which the cancer risk is so small that there is no cause for worry from a regulatory or public health perspective. EPA's pesticide program defines a risk as negligible if a person has a one-in-a-million or less chance of getting cancer as a direct result of a lifetime of exposure to a particular substance" (Environmental Protection Agency 1991b, 8). In its review process, the EPA has seldom rejected a tolerance when the risk has been less than one in a million (National Research Council 1987, 34). Because of its importance, I wish to select from the vast complexities of risk analysis—tolerance assessment, cost-benefit analysis, environmental fate, and so forth—just this position for a closer look.

This risk-threshold position is extremely important to the subject of risks associated with pesticides, both because it influences the regulatory process and because it expresses the industry view that "low-level" pesticide risks are unimportant. More specifically, when considering label directions, I described the "dose-makes-the-poison" doctrine and its role in making the case that following label directions insures pesticide safety. One of the ways of supporting that doctrine is with the threshold-of-risk position. For this reason alone it deserves a close look. However, another reason is that defenders of the industry and of the review process often assert that it is a technical matter best left to experts and that if the public could understand the process, we would be reassured by the cautions built into it. For example, Francis (1992, 1) says, "The interpretation of risk depends on who is doing the interpreting, since 'scientists' and 'consumers' interpret risks in different ways."[22]

There are at least three pitfalls to an accurate understanding of this risk-threshold position. First it must not be taken as an instance of cost-benefit analysis. On the contrary, it is a criterion for dismissing an entire category of risks prior to submitting remaining risks to cost-benefit or other analysis. Second, there is more than one sense of threshold. It can refer to no probability or to low probability. An example of the former is noise; below a certain volume, in certain situations, there is no effect. An example of low probability is any substance judged to have no greater, but also no less, than a one in one million chance of causing a specific harm. Needless to say, it is often a matter of dispute what the proper category for some substances is. And when low probability is confused as meaning no probability, then the actual risks are not being comprehended. Finally, there is a misleading vagueness to words and phrases used to characterize low-probability risks: for example, "negligible," "no cause for worry," "may be ignored," "can be disregarded," "are inconsequential," "insignificant," and so forth. What, precisely, do these words indicate? Such descriptions do not accurately convey the precise nature of the risk-threshold position. It is fundamentally an ethical judgment that in the regulatory process it is ethically permissible not to consider the full range of risk.

I turn now to problems with this position. In my discussion, I will repeatedly be drawing upon the excellent treatment of the subject by Kristin S. Shrader-Frechette (1985). In addition to these obstacles to understanding what the risk-threshold position actually means, there are also problems with the position itself. The first is the arbitrariness of the threshold. Suppose we ask why it is not set at one in five hundred thousand or one in two million. A possible reply might be that below the level of one in one million risks are no longer perceptible or meaningful. This reply will obviously not suffice, for we can think of many imperceptible albeit nontrivial risks: plutonium, radiation, or minute quantities of asbestos. One reason for placing the threshold at one in one million is that, according to Rescher (1983, 37), this approximates the chances of death by natural disaster. Apparently the assumption is that we should be willing to accept man-caused risks with similar probabilities. The ethical weakness of this rationale is aptly described by Shrader-Frechette (1985): "In assuming that an increase in risk, of the same magnitude as normal risks, is acceptable, one is really saying that it is acceptable to put certain numbers of lives in jeopardy, for whatever reason, so long as the total risk is statistically insignificant or comparable to normal rates of risk. In other words, one is assuming that a certain number of 'statistical casualties' are acceptable, even though it is widely acknowledged that the same number of predictable deaths of identifiable persons very likely would not be said to be an acceptable consequence of a given risk" (145).

A second problem with the risk-threshold position is the assumption that risk probabilities can be determined accurately. Shrader-Frechette (1985, 146–47) points out three reasons to challenge this. One is errors in modelling and measuring. For example, in assessing toxicological risk from pesticide residues, a key component is estimating what foods people eat and how often. The most recent source for this information is the 1987–88 Nationwide Food Consumption Survey compiled by the Human Nutrition Information Service, U.S. Department of Agriculture. Yet the accuracy of this survey is in serious doubt. In a report titled "Nutrition Monitoring: Mismanagement of Nutrition Survey Has Resulted in Questionable Data," the General Accounting Office (1991a) says: "Methodological problems, deviations from the survey's original design, and lax controls over the collection and processing of the results all raise doubts about the quality and usefulness of the data in the . . . survey. Most importantly, results from the survey may not be representative of the U.S. population because of low response rates" (2–3).

Another set of problems in modelling and measuring comes from appreciating that the population includes persons with special sensitivities and people who have had multiple exposures to a pesticide through sources such as residues on food and in water, application drift, and even residues in rain.[23] Someone in either situation faces a greater risk than the projected one in one million.

Still another source of measurement variability is that not all events can be accounted for in the process of estimating risk. Examples of such events are human error and foul play. In March of 1991, the EPA investigated allegations that a safety testing laboratory had fabricated research results relating to the licensing of a pesticide (Schneider 1991). Similarly, a laboratory was found to have falsified health and safety studies relating to the EPA procedures in the late 1970s and 1980s (Feldman 1991, 18). The point is not that pesticide testing is any more susceptible to fraud than any other endeavor. It is that whatever fraud there is interferes with accurate estimation of risk.

A third reason to doubt the precision with which risk can be calculated is the lack of basic scientific understanding of possible sources of risk. For example, when we are exposed to pesticide residues in mixture, the synergistic effects can be harmful to health. Yet scientists have little understanding of when and how these effects take place. One such potential source of problems today is the occurrence of atrazine and elevated levels of nitrate in ground and surface water (Keeney 1989, Hallberg 1989, and Goolsby, Coupe, and Markovchik 1991). Yet there has not been sufficient time to determine whether the agents together can have harmful consequences not associated with either separately.

In addition to the arbitrariness of the risk-threshold position and to its unacceptable claim to precision, perhaps the most serious problem with it is that it focuses exclusively on probability. As Shrader-Frechette (1985) points out, the problem with ignoring certain low probabilities is that "their magnitudes alone are not sufficient grounds for determining their acceptability. If a risk is undesirable because it would affect great numbers of persons or because in is uncompensated or unfairly distributed, then just because it is small does not mean that it is morally negligible" (142). Nuclear power is a conspicuous example. The probabilities of a catastrophic accident such as meltdown are regarded as remote. Yet such an event would be so devastating that the acceptability of nuclear power is, for that reason, continually debated. Another example is contamination of drinking water. The risks associated with unpredictable mixtures and concentration are considered not to be significant by researchers such as Goolsby, Coupe, and Markovchick (1991). Yet the potential for harm—measured in magnitude—is considerable when drinking water supplies are irreversibly contaminated. The risk-threshold position, focusing as it does only on probabilities, screens out this and other dimensions of the problem.

This discussion has set forth several major conclusions about pesticides and risk. First, pesticides fare poorly using every major criterion of acceptable risk. Second, an important and widely utilized position in risk assessment is shown to be riddled with problems. There were several reasons to closely consider the risk-threshold position. One was to see whether it is a source of reassurance about the "technicians'" employment of risk assessment in general. That reassurance has hardly been found. That could understandably arouse curiosity about other aspects of the process of risk assessment as well.

If the threshold-of-risk position is unacceptable, then a major pillar of support for the "dose-makes-the-poison" doctrine cannot stand. The consequence is that the claim of safety from following label directions—a claim which makes use of that doctrine—is revealed as even more implausible.

I recognize that raising doubts about the wisdom of dismissing low-probability risks invites unflattering characterizations and charges. Chicken Little. Nervous Nelly. Do you expect a risk-free society? Big fears over very little risks will immobilize us, stultifying our society.[24] This, however, is a false dilemma. It is not the case that we are either immobilized by concern over low-level risks or blithely ignore them. Individuals and societies can and do recognize risk types, establish and act on policies to deal with them, and then carry on with their lives. Nor does it follow from concern about low-probability risks that we do not care about other risks. Anti-smoking campaigns, AIDS research, nuclear nonproliferation, and other risk-reducing programs are not jeopardized by also seeking alternatives to toxic pesticides. I will conclude by simply returning to former Secretary of Agriculture Earl Butz's endorsement of risk-taking to cope with great challenges. If striving for an alternative to pesticide-intensive agriculture—one that retains the advantages but eliminates the disadvantages—counts as a great challenge, then it appears that by his own reasoning he would endorse the risk-taking required to move in the direction of alternative agriculture. Could that be what he means?

Conclusion

The television advertisements for pesticides continue, with their characteristic soft music and reassuring themes and images. At the same time the infrastructure of American agriculture rapidly changes to accommodate pesticide-oriented farming. The future harvest that is pesticide-free will not be upon us any time soon.

Yet this and other chapters contain ingredients of a rationale for pursuing pesticide-free agriculture. It includes the following components: First, we are not in a position to be sanguine about pesticide trends. Whether one looks more closely at the regulatory process; adopts the perspective of defensible principles of risk assessment; studies how pesticides are relocating in ground, surface, and rain water and in food; or comes to appreciate the addictive nature of pesticide-oriented farming, the situation is disturbing.

Next, pesticide-free farming is prepared, even at its inception, to effectively address a variety of agronomic challenges. These include nutrient recycling, soil conservation, efficiency, disrupting pest cycles, and, of course, the concept of sustainability. Finally, pesticide-free farming can fend off doomsday scenarios. There is no good reason to believe that such farming will lead to economic calamity, starving people, or farms overrun with livestock. Making these points will require unusual patience. Thousands of dedicated practitioners and proponents show that this ingredient, at least, is in place.

• • •

Epilogue

Why was some of the discussion of the last three chapters of this book necessary? One problem is that even though alternative agriculture is only at its inception, there are many pessimistic opinions about its prospects.

This tendency to prejudgment takes many forms. In casual conversation of both farmers and nonfarmers, the mere mention of the subject often conjures assumptions of poor performance. In agricultural academe, unfounded doomsday claims and profound skepticism are not difficult to locate. An example is found in a widely used crop production textbook: "It is estimated that a complete world reversion to organic farming would solve the population explosion problem because hundreds of millions of people would die from starvation each year" (Martin, Leonard, and Stamp 1976, 165). Another opinion is the following: "Speaking as a weed scientist, today it would be virtually impossible except on a limited basis to control weeds in crop production without the use of herbicides" (Abernathy 1990, 140). A recently published soils text asserts that "modeled projections . . . suggest that conversion to organic methods on a wide scale is likely to produce less than half as much of several important world crops" (Miller and Donahue 1990, 220). The tendency to prejudgement results repeatedly in projections of calamity, starvation, disrupted exports, financial collapse and farms overrun with livestock. It is especially disturbing to find such assertions in agricultural textbooks.

Whatever the reasons for this continual disparagement, the discussion does not have to proceed in such a manner. Why rely upon modeled projections, assumptions, and conjecture? There are several reliable sources of data and insight. One is examination of actual alternative farms scattered across North

America and other regions of the world. Investigation could begin with a more systematic conception of what can and cannot be learned from actual systems. Another source of information is research. It is hard to imagine a claim on behalf of alternative agriculture that is not susceptible to empirical investigation. Happily, this endeavor has now begun across the continent. It is taking place at universities, the U.S. Department of Agriculture, and numerous other institutions. While in the view of some of us this research commitment is much delayed, that simply provides incentive to make up for lost time. Yet another touchstone is pure deductive analysis. For example, some discussions about the feasibility of alternative agriculture set forth assumptions which exclude intrinsic components of that agriculture, such as cropping diversity or livestock. Reasoning alone can disclose that such projections are misconceived.

The abiding question concerns the long-term viability of pesticide-free, conservation-oriented agriculture. I propose that we pursue the question on its merits, with only the honest use of the best intellectual tools at our disposal.

• • •

Appendix
Profile of the Bender Farm

Our 642-acre farm is located in Cass County, Nebraska, between Lincoln and Omaha, just a few miles from the Platte and Missouri rivers on the western edge of the dryland Corn Belt. The land is in four separate tracts. On the farms we raise 12 crops in rotation without irrigation: corn, milo, soybeans, wheat, oats, forage sorghum, alfalfa, sweet and red clover, prairie hay, rye, and turnips (for forage). The cropland has been free of herbicides, insecticides, and fungicides since 1980, and it has received only alternatives to synthetic fertilizers since 1987. The field crops of the farm were certified organic by the Organic Crop Improvement Association of America in 1990, 1991, and 1992. A wide array of conservation structures are in place and conservation practices followed.

Climate

In the winter, Cass County is usually cold; the average temperature is 26 degrees Fahrenheit. Summers are usually hot, though there are occasional cool periods. The average summer temperature is 75 degrees, with an average daily maximum of 87 degrees. The growing season is 145 days nine years out of ten.

Average yearly precipitation is 32 inches, 24 of which usually fall April through September. In two years out of ten the rainfall during this summer period drops to 18 inches. Thunderstorms occur on about 50 days of the year; they are often intense, with large amounts of rain falling in a short time. Average yearly snowfall is 29 inches. Average relative humidity is 60% in midafternoon and 80% at dawn (Borchers et al. 1984, 2).

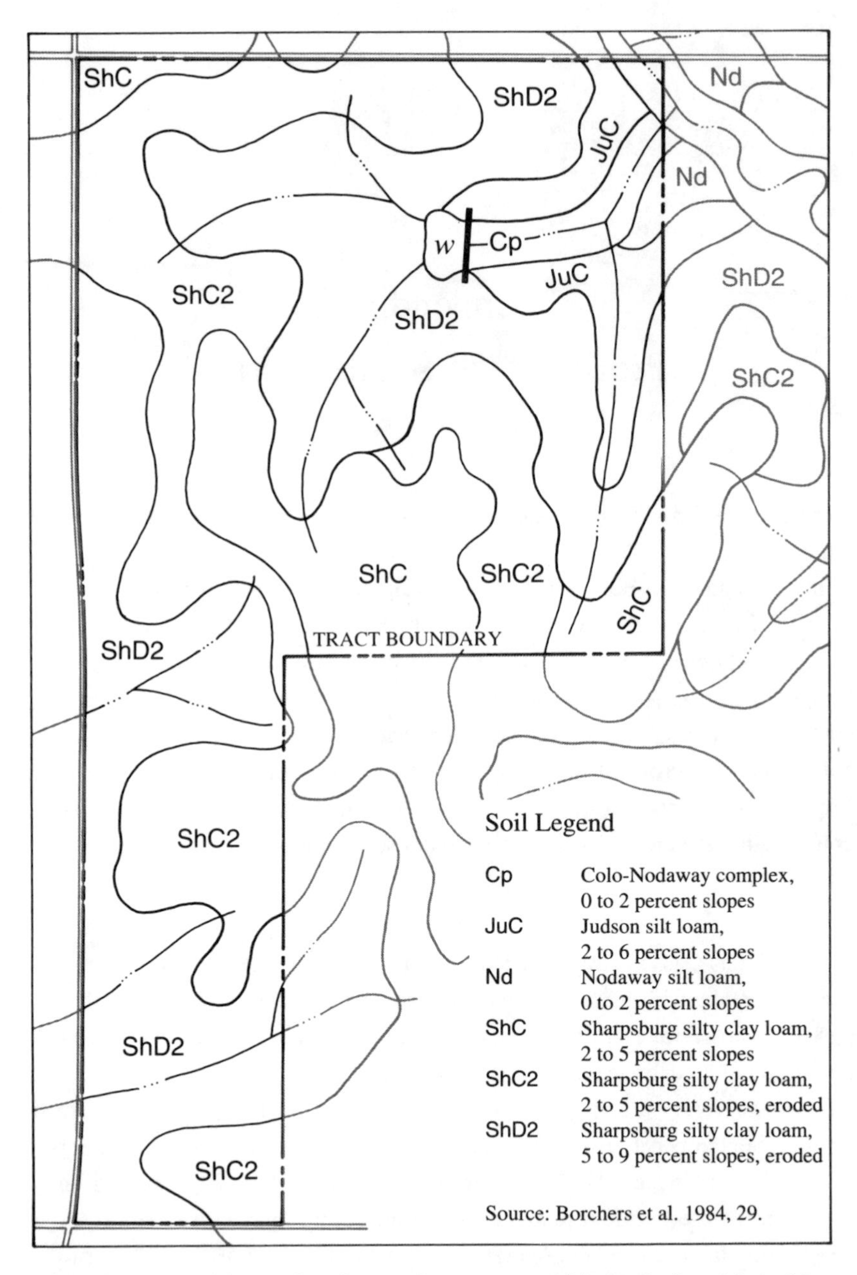

Map 1. Soil types and degree of previous erosion on one tract of the Bender farm. Much of the erosion took place in the first part of this century, prior to the advent of modern soil conservation techniques.

Table 5. Yields per acre projected by the Cass County Soil Survey for two soil/slope types present on the Bender farm

	corn	soybeans	milo	winter wheat	alfalfa hay
Sharpsburg Sh (0–2% slope)	98 bu	40 bu	98 bu	44 bu	4.8 tons
Sharpsburg ShD2 (5–9% slope, eroded)	80 bu	32 bu	82 bu	37 bu	3.7 tons

Source: Based on Borchers et al. 1984, 102.

Soil and Terrain

According to the *Soil Survey of Cass County, Nebraska* (Borchers et al. 1984), almost all soils on the four tracts are Sharpsburg silty clay loam. These are generally deep, moderately well-drained soils formed in loess on uplands. Permeability is moderately slow. A few acres include Judson soils, which have a thicker surface horizon and are on foot slopes below the Sharpsburg soils. There is also some Nodaway silt loam, Colo-Nodaway complex, and Colo silty clay loam.

The *Soil Survey of Cass County, Nebraska* distinguishes Sharpsburg soils according to slope and erosion. Map 1, from this survey, includes one of our farm tracts.

Notice that a considerable percentage of the tract is identified as ShD2, which indicates 5 to 9% slopes and eroded. This tract is representative of the other three in this respect. The *Soil Survey* says of lands designated ShD2 that "erosion has removed much of the original dark surface soil over most of the area. . . . Runoff is rapid. Organic matter content is moderately low, and natural fertility is medium"(42). Severe erosion on such soils was reported already in the 1941 *Soil Survey of Cass County, Nebraska* (Beesley et al. 1941, 26). All the acreage of three tracts and most of the fourth has been designated as potentially erodible by the U.S. Soil Conservation Service.

Table 5 presents yields per acre projected by the 1984 *Soil Survey* according to soil type and slope.

Facilities

Three of the four tracts have sets of buildings. They include old-style hay and horse barns, wooden corn cribs, and granaries. Some have been converted to store hay, grain, and machinery. In addition, there are three modern grain bins which hold a total of 16,000 bushels and a large, modern building for storing hay and machinery. Three of the four tracts have complete cattle handling facilities. Each of the four tracts is fenced, including either pasture fence or fenced-in fields.

Machinery

On our farm we have six regularly-used tractors, four of which were purchased used, ranging in horsepower from 60 to 160. The oldest is a 1968 model and the newest a 1976. The total hours of use for all six tractors is 27,000. The combine is a rotary model purchased in 1980. We have no large truck, but numerous wagons, including three large grain carts, several smaller barge wagons, and many wagons and trailers for baled hay and straw that together can hold over 1,500 small square bales at one time. Tillage equipment includes an offset disk (purchased used), a winged field cultivator, and a moldboard plow. We have several old (1950s) grain drills and a 12-row planter. Weed cultivation equipment includes a 31-foot rotary hoe, a 26-foot spring-tine harrow, an old (1960s) 6-row crop cultivator (purchased used), and a chisel plow with sweeps for noxious perennials (purchased used). Custom operators are hired to windrow hay, stack hay, grind feed, and haul cattle.

This particular set of machinery should not be considered as entirely necessary for such an operation. From 1980 through 1983 we farmed all 642 acres without pesticides using aging 4-row crop equipment and three modern tractors of 60, 90, and 125 horsepower.

Livestock

On our farm we have 90 to 100 beef cows, with cattle on all four tracts most of the year. Some calves are marketed at weaning, some after backgrounding, and some at finish. Most of the replacement heifers are raised on the farm. The cattle do a considerable amount of foraging for feed; this is made possible by permanent fences, diversity of cropping, absence of fall tillage, and the planting of turnips as a second crop.[1]

We have not sought organic certification of the livestock because we use neighbors' fields for winter grazing and because we use ivomectin for parasites. Ivomectin is controversial, and debate continues as to whether or not its use should be approved by the Organic Crop Improvement Association. Individual animals are medicated as needed.

Crop Management

The crop rotation has some flexibility. It commences with a soil-building crop: alfalfa, sweet clover, or red clover. The next crop is a feed grain, either corn or grain sorghum. The feed grain is alternated with soybeans once or twice. After soybeans comes oats or wheat. Then follows turnips, alfalfa, clovers, or one more cycle of soybeans. A row crop is planted after turnips. Sometimes soybeans are planted two years in a row, and sometimes forage sorghum follows soybeans.

The objectives are to alternate sod-base crops with row crops, weed suppressing crops with those without that characteristic, crops susceptible to specific insects with those that are not, and soil enhancing crops with those that do not enhance soils.

The annual soil tests generally reveal high levels of potassium and moderate to low levels of phosphorus. Trace minerals are generally adequate, except for an occasional slight deficiency in zinc. The soil pH is about 7.0. One of the most important methods of managing soil fertility is careful crop rotation. Additional sources of fertility include legumes, green manure, cattle manure (generally spread in spring, summer, and fall), North Carolina rock phosphate (which has much higher availability than typical rock phosphate), lime (from local quarries), and miscellaneous biomass, such as woodchips, ashes, and large quantities of leaves from the nearby community.

For tilling fields with heavy residue, an offset disk is used. Otherwise a field cultivator is the tool of choice, sometimes even in cornstalks after heavy grazing. A moldboard plow is used to till alfalfa fields, to work stubborn weed problems such as hemp dogbane and morning glory, and to reinforce terraces.

We plant row crops with a surface planter. Ridge-till is not feasible on our farm because almost all of the land is terraced.

An important component of our weed control strategy is carefully planned crop rotation. Among its benefits for weed control are that it interrupts weed cycles, it spreads out the weed control work, it makes possible diversified tillage, and it creates allelopathic opportunities. Other aspects of weed control are planting at specific dates, but allowing sufficiently long intervals between fields, and planting when soil moisture is correct. Mechanical weed control includes using a rotary hoe, a mounted spring-tine harrow, and a shovel cultivator. We do hand roguing as needed.

Our strategy for insect control is to use careful crop rotation and crop placement, to limit field size, and to use seed varieties with insect resistance.

Soil Conservation

On our farm we use a variety of soil conservation practices. Crops are diversified and in rotation. The emphasis is on legumes and sod-base crops; only about 30 to 35% of the total acres are usually in row crops. We avoid early spring tillage (with the exception of oats planting) and fall tillage. Wherever

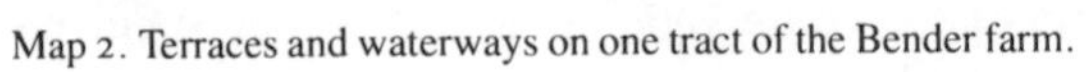

Map 2. Terraces and waterways on one tract of the Bender farm.

possible, we till with the goal of leaving high residue on the surface. Much of the planting is done as strip crops and on contours. We have over 30 miles of water diversion terraces, some of them very large ones with 30-foot front and back slopes. Map 2 shows the terraces on one of the farm tracts.

Over 25 acres are in grassed waterways; almost all are over 40 feet wide, and some are over 60. In addition, many acres of highly erodible land have been seeded to permanent grass, and we have planted many trees.

Performance Indicators

These soil conservation practices seem to be working well. Although the usual period of effectiveness for terraces and waterways is 10 years, ours have lasted much longer. For example, several waterways in areas with 5 to 9 degree slopes are 20 to 35 years old and have retained their structural integrity without repair. Even after severe thunderstorms, such as one in 1990 in which 8 inches of rain fell in a 10-hour period, the conservation structures have functioned as designed, sustaining only minimal damage.

Weed and insect management have also been successful. During the many farm tours that stop at our farm, weed management is one of the things that is highlighted. Even in very wet years, such as 1984 and 1990, the weed situation was manageable. And sometimes we have obtained excellent results with very little mechanical cultivation and no hand roguing. Only minimal insect damage has occurred in the last 11 years. Occasionally there are weevils in alfalfa and greenbugs in grain sorghum, and recently we have had some chinch bug damage at field borders in grain sorghum and forage sorghum. However none of these have been serious.

Our crop yields have been competitive for our area, despite the fact that much of our cropland is ShD2, that is, 5 to 9% slope and significantly eroded. In years of minimally adequate (or greater) rainfall, our yields have exceeded the alfalfa yields and matched the corn, soybean, and milo yields projected by the *Soil Survey of Cass County, Nebraska* (Borchers et al. 1984, 102) for the Sharpsburg Sh soil/slope type, that is, 0 to 2% slope (see table 5). Our wheat yields are closer to yields projected for the Sharpsburg ShD2 soil/slope type, perhaps affected adversely by our heavy fall and light spring pasturing.

Profitability is determined not only by yields, which are comparable to those from conventional farms in our area, but also by two other aspects of an operation such as ours: costs and lower-profit crops such as wheat and oats. The costs of weed control in row crops vary, depending on the amount of rainfall during the planting season.

Row-crop weed-control costs in dry years	
1 rotary hoeing	$3 per acre
1 cultivation	$3 per acre
minimal hand roguing	$2 to $5 per acre

The exception is grain sorghum. It requires hand roguing of shattercane each year at $4 to $7 per acre. However, this is a problem which is not peculiar to nonpesticide weed control in sorghum.

Row-crop weed-control costs in wet years	
2 rotary hoeings	$6 per acre total
1 harrowing or 1 additional hoeing	$3 per acre
2 cultivations	$6 per acre total
extensive hand roguing	$10 to $20 per acre

An additional cost in wet years is the occasional need to replant, which occurs once every 10 to 12 years on average per field. In years of ordinary rainfall, costs range between these two extremes, depending upon how rain falls in relation to planting dates.

In addition to the proceeds from the sale of the grain, the value of the wheat and oats crops is enhanced in several ways. We use wheat fields for fall and spring pasturing,[2] and we use the straw for bedding and as a feed supplement. After the wheat fields are harvested, we plant turnips for forage. The value of the oats crop is increased by using oat hay and straw for feed, by double cropping to turnips, and by feeding the grain to calves, replacement heifers, and bulls and using it in the finishing ration. In addition, we sell seed oats which are registered and/or certified and certified organic.

The revenues of the farm benefit from no special markets other than that for seed oats prior to 1993. The total commodities sold as organic since 1980 have been only 10 to 15 fat steers and 300 to 350 bushels of soybeans.

Problems and Challenges

As one would expect in any farming operation, we have a number of continuing problems and challenges, leaving considerable room for improvement. One goal is to enhance the management of grazing in order to reduce pasture weed problems. A second challenge, one which will require considerable effort, is to restore the soil on slopes that were severely eroded in the first part of the century. This challenge would be much the same if the farm were managed conventionally. A third area for improvement is in controlling livestock parasites with nonchemical means. A fourth problem—one not special to nonchemical agriculture—is perennial noxious weeds. I am experimenting to find more effective nonchemical methods. And finally, a persistent problem is to provide sufficient nitrogen for wheat and forage sorghum in the rotation.

• • •

Notes

Preface

1. Fifteen acres of wheat were mistakenly sprayed with 2,4-D by a neighbor in 1985, the result of family miscommunication.

Chapter 1

CHANGE: GOALS AND OBSTACLES

1. Martin, Leonard, and Stamp (1976, 69 and 312) describe the properties of bindweed which make this characteristic possible. Rosenthal (1983) gives an account of the rapid spread of bindweed between 1965 and 1980. According to Peters (1991, 9), the lack of effectiveness of herbicides on bindweed has been a frustration at the Rodale Research Center's Farming Systems Trial.

2. See also Francis and Youngblood 1990, 8–10.

3. See Jukes 1990, 31–32; Watson 1990, 92–93; Marten 1990, 119; and Beus and Dunlap 1990.

4. See also Hallberg 1989 and Environmental Protection Agency 1986, 3–9.

5. See also Aspelin, Grube, and Kilber 1991 and U.S. Department of Commerce 1987, 19.

Chapter 2

CONVERSION

1. This problem probably resulted either from a lack of sufficient alfalfa or from insufficiently diversified tillage. See Miller and Donahue 1990, 437 for a discussion of a relationship between reduced tillage and promoting rhizomatous grasses and hemp dogbane.

2. For a related discussion see Andrews, Peters, and Janke 1990.

3. For similar conclusions about runoff and erosion, see Saxton, Spomer, and Kramer 1971. For another discussion of the positive effects of terraces on retaining moisture, see Miller and Donahue 1990, 150–52.

4. For an elaboration on several of these advantages of alfalfa, see Martin, Leonard, and Stamp 1976, 621–40. For a discussion of the advisability of beginning a long rotation with a legume, see Duffy 1991, 97–98.

5. Enthusiasm about oats can be considerable. The great writer Anton Chekhov, who was also a physician, is described by his biographer Troyat as having related the following incident in a letter:

> A peasant woman who had been seriously wounded in a fall from a cart interspersed moans about her imminent death with instructions to her husband about the oat harvest. "I told her to forget about the oats," he wrote to Suvorin, "we had more serious things to talk about, and she said, " 'But he's got such good oats!' I envy her spirit. People like that have no trouble dying." (Troyat 1986, 142)

6. For relevant descriptions of sorghum, see Martin, Leonard, and Stamp 1976, 383–400.

7. For the lime requirements of selected crops, see Miller and Donahue 1990, 241.

8. For information on nematodes, see Martin, Leonard, and Stamp 1976, 709.

Chapter 3

WEED MANAGEMENT

1. For a discussion of the soil enhancing characteristics of alfalfa and grasses, see Brady 1974, 59. For the degranulating effects of corn, see Brady 1974, 63.

2. For a discussion of rotating to increased tillage in reduced tillage situations, see Miller and Donahue 1990, 438–39.

3. For a discussion of the negative effect of tillage, see Brady 1974, 60.

4. One calculation is that winter weathering results in a 10 to 30% decrease in surface residue (Dickey, Jasa, and Shelton 1986, table 1).

5. For a detailed description of power requirements for implements, see Hunt 1976, 40, table 2.4.

6. For details about early germinating weeds and the implications for weed control, see Gunsolus 1990, 114–19.

7. See also Pfeiffer 1987, Raymor and Bernard 1988, and Egli, Guffy, and Heitholt 1987.

8. For additional reasons to cultivate early, see Schauer 1991, 25.

9. For statistics on the spread of bindweed in California between 1965 and 1981, see Rosenthal 1983, 16.

Chapter 4

LIVESTOCK

1. The relationship between tax policy and livestock production is explained in detail in Strange 1988, chapter 7.

2. For example, see Singer 1990.

3. For an excellent discussion of the importance of livestock to alternative agriculture, see also Coppinger, Clemence, and Coppinger 1992, 21–23.

4. What value is this point to absentee farm owners who usually do not share ownership of livestock? Livestock enable the tenant to pay pasture rent on otherwise wasted acres, enhance soil resources, and steadily reduce costs for fertilizers and pesticides. If this is a problem, however, especially in cash rent situations, it is a problem for separation of ownership and operation of farms, *not* organic farming.

5. In a few years, erodible acres which were seeded to grass as part of the federal long-term Conservation Reserve Program will no longer be in that project. The chances are slim that these fields will remain in grass on farms where there is no livestock. This could be tragic, given the general difficulty of getting farmers to practice soil conservation. The 1982 National Resources Inventory conducted by the Soil Conservation Service discloses that entire regions of the United States are losing soil at the rate of 15 to 25 tons per acre per year. Agriculture can hardly afford any additional disincentives to soil conservation.

6. For a detailed discussion of the vital role of livestock in achieving plant-nutrient balance in Swedish agriculture, see Granstedt 1992.

7. For example, see the criticism of Palmer (quoted in Logsdon 1986, 8–9).

8. For an insightful discussion of these relationships, see Coppinger, Clemence, and Coppinger 1992, 21–23.

9. For a detailed discussion of problems for Swedish agriculture as a result of the separation of livestock from farms, see Granstedt 1992.

10. This model would not be suited to farms where marginal land is predominant. Such farms generally require more livestock. I should stress that this exploration is intended to be an illustration, not a prescription. Obviously practitioners of organic farming are meeting their objectives with

other kinds and quantities of livestock. I have selected beef cattle merely to illustrate a point.

11. One study concludes that yields of a grain crop grown in rotation are 10 to 20% greater than those in continuous grain, regardless of fertilizer (Langer and Randall 1981, 182).

12. See also Singer 1990.

13. For elaboration on these possibilities, see Klopfenstein 1991. For elaboration on the forage potential and other possibilities for livestock in sustainable systems, see Benbrook 1991.

14. I have visited with persons so hostile to the idea of livestock that when confronted with the relationship to organic farming, they are tempted to proclaim, "So much the worse for organic farming!" One problem with that reaction is that a central incentive for organic farming is its positive impact upon fish and wildlife. See Robinson 1990.

15. For more information on the dry lot feeding concept, see Anderson and Boyles 1989 and Neumann 1977, 300–304.

Chapter 5

COMPARING SYSTEMS

1. Chemical no-till agriculture is not plagued with every problem discussed here. I will deal with no-till from other perspectives in the next section.

2. There is some evidence that certain aspects of organic farming inherently minimize soil erosion. Cacek (1984) describes work done in the first part of the century on the relationship between crop production and soil erosion. While it is hardly surprising that crop rotation was shown to conserve soil, the results went beyond that: "The corn years on the rotation plot were compared with the same years on the continuous corn plot. . . . Erosion on the continuous corn plot was 4.7 times greater than that on the rotational corn, suggesting that ground cover alone did not explain the lower erosion rate on the rotation plot. Presumably changes in soil structure influence soil erodibility" (Cacek 1984, 358). Another experiment reported "a soil loss of 7.1 tons per acre for corn following corn but only 4.2 tons per acre for corn following clover and 2.4 tons per acre for corn following two years of grass-clover hay. Soil loss for soybeans following corn was 6.9 tons per acre versus only 2.0 tons per acre following one year of grass-clover hay. These data show that legume-based crop rotations significantly reduce soil erosion" (358). A comparison of soil properties of two farms in the Palouse, one managed organically and one conventionally for de-

cades, found dramatically less soil erosion on the organic farm. According to Reganold (1988), "the USLE [universal soil loss equation] erosion estimate for the conventionally-farmed soil is 2.6 times that for the organically-farmed soil" (151).

3. For the observation that high amounts of organic matter on the soil surface impede normal functioning of herbicides, see Miller and Donahue 1990, 437. See also Monsanto encapsulation videotape no. 114–91–11.

4. For a discussion of the effectiveness of various cover crops, see Miller and Donahue 1990, 458.

5. For a description of twelve disadvantages of no-till, see Miller and Donahue 1990, 434–36.

6. For a discussion of the causal factors, see Miller and Donahue 1990, 58 and Brady 1974, 50–51.

7. Soil scientists debate precisely how much and for how long legumes and tillage reduce compaction. For the view that deep tillage does little in the long term and that a more effective approach is a long rotation which includes legumes, see Smith 1992.

8. Of course, the organic farmer's hay handling equipment must be taken into account in the analysis. However, a mere one-step increase in combine capacity alone can cost as much as or more than a hay baler.

9. For an attempt to quantify the savings of this kind of management, see Duffy 1991.

10. For an analysis of financial stability in farming systems, see Helmers, Langemeier, and Atwood 1986.

11. See Strange 1988, chapter 6, for a discussion of inflexibility and risk as they apply to big versus small farms and the implications for farm programs.

12. For an extended discussion of economic evaluation of alternative farming systems, see National Research Council 1989, chapter 4.

13. For a discussion of pitfalls of cover crop experiments at the Thompson farm in Iowa, see Schauer 1991, 13–17.

14. For a discussion of other issues and problems with studies of actual farms, see Cacek and Langner 1986, 25–26.

15. For another description of contrasting work loads see Exner, Thompson, and Thompson 1990.

16. For another description of the uses of multiple tractors on a pesticide-free farm, see National Research Council 1989, 258.

Chapter 6

THE ASSAULT ON ALTERNATIVE AGRICULTURE

1. Not all studies of this subject are pessimistic. See Pimental 1991 and Helmers, Azzeddine, and Spilker 1990.

2. An argument with similar logic points out that aspects of alternative systems can be mismanaged. See Council for Agricultural Science and Technology 1990, 93. The problem with this argument also is that any human endeavor—education, medicine, and so forth—can be mismanaged. Such an argument needs additional premises which show that the enterprise in question displays inherent susceptibility to problems of mismanagement. That is why the determination that ground water contamination by pesticides is associated with normal agricultural practice is so devastating.

3. Some regard this study as extreme insofar as it includes scenarios of no chemical use whatsoever. I, understandably, choose to measure the extremeness of a study by the duration and method of its projected conversion. See, for example, Schaller 1991.

4. For examples of similar studies, but ones using a variety of methods and reaching a variety of conclusions, see Olson, Langley, and Heady 1982; Pimental 1991; Helmers, Azzeddine, and Spilker 1990; Council for Agricultural Science and Technology 1980; and U.S. Department of Agriculture 1980. See also Madden 1990 for a discussion of the issues that must be considered in thinking about the economics of alternative farming systems.

5. Variations of the sentence are repeated on pages 2 and 43 and again in a response to critics in Knutson et al. 1990b, 31.

6. For a discussion of the idea that alternative agriculture requires physical *and* conceptual changes, see Coleman 1989, 169.

7. For criticism of systems studies, see Aldrich 1990, 67.

8. For related discussion, see National Research Council 1989, 247–418, Dobbs et al. 1991, and Thompson and Thompson 1991.

9. See Sutter 1991.

10. On the erodibility of corn ground in differing residue and rotation situations, see Miller and Donahue 1990, 457. For a cautionary discussion about corn and soil erosion, see Martin, Leonard, and Stamp 1976, 344.

11. On the tendency of corn to degranulate soils, see Brady 1974, 63.

12. If "normal agricultural use" refers to something other than following label directions, we then are confronted by a different problem.

13. See also Hallberg 1989 and Environmental Protection Agency 1986, 3–9.

14. For more details see National Research Council 1987.

15. The efficiency loss computation was also very controversial. See Hinkle and Pace 1987, 15–24. For a general discussion of these issues, see Osteen and Szmedra 1989, 48, and General Accounting Office 1991b.

16. For a similar argument, see Warnock 1987.

17. For detailed elaboration of this general idea, see Green 1977, 260–65.

18. See also Francis 1992.

19. See Lowrance 1976, 79–94, and American Chemical Society 1984, 10, for general discussion of basic guides to judging risks.

20. For a detailed discussion of circumstances in which occupational hazards might be ethically justified, see chapter 4 of Shrader-Frechette 1985.

21. National Research Council 1987, 29. Other tests and data are required for the categories of chemistry, environmental fate, and ecological effects (Environmental Protection Agency 1991b, 3).

22. For three assertions that regulatory procedures are designed to overstate risks and, hence, err on the side of caution, see Rodricks and Taylor 1990; National Research Council 1987, especially chapter 2; and Environmental Protection Agency 1991b, 7–9.

23. EPA itself acknowledges this issue (Environmental Protection Agency 1991b, 9). For recent accounts of the emerging phenomenon of pesticide rain, see Thomas 1991, 9, 11 and Associated Press 1990, 21.

24. See for example Resher 1983, 40.

Appendix

PROFILE OF THE BENDER FARM

1. For a discussion of the general advantages of high-forage beef production and of its suitability for a region which includes eastern Nebraska, see Klopfenstein 1991.

2. For an analysis of the value of wheat as winter pasture, see Neumann 1977, 263.

• • •

References

Abernathy, John R. 1990. Alternative agriculture: Whose perspective? In *Alternative agriculture scientists' review,* 139–40. Council for Agricultural Science and Technology Special Publication no.16. Ames, Ia.

Agricultural Council of America. 1990. FoodWatch for tomorrow: Balancing needs and resources. Washington, D.C.: ACA Education Foundation.

Aldrich, Samuel R. 1990. Alternative agriculture: A timely, but flawed report. In *Alternative agriculture scientists' review,* 66–68. Council for Agricultural Science and Technology Special Publication no.16. Ames, Ia.

American Chemical Society. 1984. *Chemical risk: A primer.* Washington, D.C.: American Chemical Society, Department of Government Relations and Science Policy.

Ames, Bruce N., Renae Magaw, and Lois S. Gold. 1990. Ranking possible carcinogenic hazards. In *Readings in risk,* ed. Theodore S. Glickman and Michael Gough, 76–92. Washington, D.C.: Resources for the Future.

Anderson, V. L., and S. L. Boyles. 1989. *Drylot beef cow/calf production.* Fargo: North Dakota State University Extension Service.

Andrews, Rebecca W., Steven E. Peters, and Rhonda R. Janke. 1990. Converting to sustainable farming systems. In *Sustainable agriculture in temperate zones,* eds. Charles A. Francis, Cornelia B. Flora, and Larry D. King, 281–313. New York: John Wiley and Sons.

Aspelin, Arnold L., Arthur H. Grube, and Virginia Kilber. 1991. *Pesticide industry sales and usage: 1989 market estimates.* Washington, D.C.: Office of Pesticide Programs, Environmental Protection Agency.

Associated Press. 1990. Atrazine rain spreading from Corn Belt, *Omaha World Herald,* 18 Dec., 21.

Baum, Warren C. and Michael L. Lejeune. 1986. *Partners against hunger*. Washington, D.C.: The World Bank.

Beesley, T. E., W. S. Gillam, M. N. Kuper, and T. K. Popov. 1941. *Soil survey of Cass County, Nebraska*. Washington, D.C.: Bureau of Plant Industry, U.S. Dept. of Agriculture.

Benbrook, Charles M. 1991. *Sustainable agriculture in the 21st century: Will the grass be greener?* Washington, D.C.: The Humane Society of the United States.

Beus, Curtis E. and Riley E. Dunlap. 1990. Conventional versus alternative agriculture: The paradigmatic roots of the debate. *Rural Sociology* 55: 590–616.

Borchers, Glenn A., Doug Witte, Steve Hartung, and John D. Overing. 1984. *Soil survey of Cass County, Nebraska*. N.p.: U.S. Dept. of Agriculture, Soil Conservation Service.

Brady, Nyle C. 1974. *The nature and properties of soils*. 3d ed. New York: Macmillan.

Butel, F. H., and Gilbert W. Gillespie. 1988. *Agricultural research and development and the appropriation of progressive symbols: Some observations on the politics of ecological agriculture*. Cornell University Department of Rural Sociology Bulletin no.151.

Butz, Earl. 1989. Guest editorial. *Crop Protection Management,* Feb., 3.

Cacek, Terry. 1984. Organic farming: The other conservation farming system. *The Journal of Soil and Water Conservation* 39: 357–60.

Cacek, Terry, and Linda L. Langner. 1986. The economic implications of organic farming. *American Journal of Alternative Agriculture* 1: 25–29.

Call, L. E. and R. E. Getty. 1923. *The eradication of bindweed*. Kansas Agricultural Experiment Station Circular no.101.

Carson, Rachael. 1962. *Silent spring*. Boston: Houghton Mifflin.

Coleman, Eliot. 1989. *The new organic grower*. Chelsea, Vt.: Chelsea Green.

Coppinger, Raymond, Elisabeth Clemence, and Timothy Coppinger. 1992. The role of livestock in sustainable agriculture. *The Land Report,* no.43: 21–23.

Council for Agricultural Science and Technology. 1980. *Organic and conventional farming compared*. Report no.84. Ames, Ia.

———. 1990. *Alternative agriculture scientists' review*. Special Publication no.16. Ames, Ia.

Dickey, Elbert C., Paul J. Jasa, and David P. Shelton. 1986. Estimating residue cover. *NebGuide* no. G86–793. Lincoln, Nebr.: Cooperative Extension Service, Univ. of Nebraska.

Dobbs, Thomas L., James D. Smolik, and Clarence Mends. 1991. On-farm research comparing conventional and low-input sustainable agriculture systems in the northern Great Plains. In *Sustainable agriculture research and education in the field*, 250–65. Washington, D.C.: National Academy Press.

Drèze, Jean, and Amartya Sen. 1989. *Hunger and public action*. Oxford: Clarendon Press.

Duffy, Michael. 1991. Economic considerations in sustainable agriculture for Midwestern farmers. In *Sustainable agriculture research and education in the field*, 92–106. Washington, D.C.: National Academy Press.

Effertz, Nita. 1985. Trumping turnip troubles. *Farm Journal*, Oct., 20–21.

Egli, D. B., R. D. Guffy, and J. J. Heitholt. 1987. Factors associated with reduced yields of delayed plantings of soybeans. *Journal of Agronomy and Crop Science* 159: 176–85.

Elmore, Roger W., and A. Dale Flowerday. 1984. Soybean planting date: When and why. *NebGuide* no. G84–687. Lincoln, Nebr.: Cooperative Extension Service, Univ. of Nebraska.

Environmental Protection Agency. 1986. *Pesticides in ground water: Background document*. WH-550G. Washington, D.C.: U.S. Government Printing Office.

———. 1990. *Suspended, cancelled, and restricted pesticides*. EN-342. Washington, D.C.: U.S. Government Printing Office.

———. 1991a. Ethyl parathion, receipt of requests for cancellation, maintenance fee cancellation, cancellation order, notification requirement, memorandum of agreement, request for comment on tolerance reduction/revocation. *U.S. Federal Register* 56, no.240 (Dec. 13) 65061.

———. 1991b. *EPA's pesticide programs*. H-7506-C. Washington, D.C.: U.S. Government Printing Office.

———. 1991c. *Pesticide reregistration progress report*. H-7508W. Washington, D.C.: U.S. Government Printing Office.

Essa, Talib A. 1979. Influence of planting date on yield, dry matter accumulation, and morphological characteristics of six soybean cultivars *Glycine max* (L.) Merrill. Master's thesis, University of Nebraska–Lincoln.

Exner, Derrick, Richard Thompson, and Sharon Thompson. 1990. Case study: A resource-efficient farm with livestock. In *Sustainable agriculture in temperate zones,* eds. Charles A. Francis, Cornelia B. Flora, and Larry D. King, 263–80. New York: John Wiley and Sons.

Feldman, Jay. 1991. Statement of Jay Feldman, National Coordinator, National Coalition against the Misuse of Pesticides before the Government Activities and Transportation Subcommittee, Committee on Government Operations, U.S. House of Representatives, 3 Oct.

Foster, G. R., and R. E. Highfill. 1983. Effect of terraces on soil loss: USLE P factor for terraces. *Journal of Soil and Water Conservation* 39: 48–51.

Francis, Charles A., Cornelia B. Flora, and Larry D. King, eds. 1990. *Sustainable agriculture in temperate zones*. New York: John Wiley and Sons.

Francis, Charles A., and Garth Youngberg. 1990. Sustainable agriculture: An overview. In *Sustainable agriculture in temperate zones,* eds. Charles A. Francis, Cornelia B. Flora, and Larry D. King, 1–23. New York: John Wiley and Sons.

Francis, F. J. 1992. *Food safety: The interpretation of risk*. No. CC1992-1. Ames, Ia.: Council for Agriculture Science and Technology.

General Accounting Office. 1986. *Pesticides: EPA's formidable task to assess and regulate their risks*. GAO/RCED-86-125. Washington, D.C.: U.S. Government Printing Office.

———. 1989. *Reregistration and tolerance reassessment remain incomplete for most pesticides*. T-RCED-89-40. Washington, D.C.: U.S. Government Printing Office.

———. 1991a. *Nutrition monitoring: Mismanagement of nutrition survey has resulted in questionable data*. GAO/RECD-91-117. Washington, D.C.: U.S. Government Printing Office.

———. 1991b. *Pesticides: Better data can improve the usefulness of EPA's benefit assessments*. GAO/RCED-92-32. Washington, D.C.: U.S. Government Printing Office.

———. 1991c. EPA's repeat emergency exemptions may provide potential for abuse. Statement of Peter F. Guerrero, Associate Director, Environmental Protection Issues, before the Subcommittee on Environment, Committee on Science, Space, and Technology, U.S. House of Representatives, 23 July.

Goldstein, Alan H. 1986. Bacterial solubilization of mineral phosphates: Historical perspective and future prospects. *American Journal of Alternative Agriculture* 1: 51–57.

Goolsby, R. C., R. C. Coupe, and D. J. Markovchick. 1991. Distribution of selected herbicides and nitrate in the Mississippi River and its major tributaries, April through June 1991. Denver: U.S. Geological Survey, Water Resources Investigations Report 91-4163.

Granstedt, Artur. 1991. The potential for Swedish farms to eliminate the use of artificial fertilizers. *American Journal of Alternative Agriculture*. 6: 122–31.

Green, Ronald M. 1977. Intergenerational distributive justice and environmental responsibility. *BioScience* 27: 260–65.

Gunsolus, Jeffry L. 1990. Mechanical and cultural weed control in corn and soybeans. *American Journal of Alternative Agriculture*. 5: 114–19.

Hallberg, G. R. 1989. Pesticide pollution of groundwater in the humid United States. *Agriculture, Ecosystems and Environment* 26: 299–367.

Hanson, N. S., F. D. Keim, and D. L. Gross. 1943. *Bindweed eradication in Nebraska*. Rev. ed. Nebraska Agricultural Experiment Station Circular no.50.

Hattis, Dale, and David Kennedy. 1990. Assessing risks from health hazards: An imperfect science. In *Readings in risk,* ed. Theodore S. Glickman and Michael Gough, 156–63. Washington, D.C.: Resources for the Future.

Helmers, Glenn A., Michael R. Langemeier, and Joseph Atwood. 1986. An economic analysis of alternative cropping systems for east-central Nebraska. *American Journal of Alternative Agriculture* 1: 153–58.

Helmers, Glenn A., Azzam Azzeddine, and Matthew F. Spilker. 1990. U.S. agriculture under fertilizer and chemical restrictions. Part 1. Lincoln: University of Nebraska Department of Agricultural Economics Report no.163.

Hinckle, Maureen K. and Charles E. Pace. 1987. Comments on EPA's proposed decision to reregister alachlor. Washington, D.C.: National Audubon Society.

Hunt, Donnell. 1976. *Farm power and machinery management*. 6th ed. Ames: Iowa State University Press.

Johnson, Douglas R. 1979. Influence of planting date on yield, dry matter accumulation, nutrient uptake and several agronomic characteristics on six soybean varieties *Glycine max* (L.) Merrill. Master's thesis, University of Nebraska–Lincoln.

Jukes, Thomas H. 1990. Comments on alternative agriculture. In *Alternative agriculture scientists' review,* 28–32. Council for Agricultural Science and Technology Special Publication no.16. Ames, Ia.

Keeney, D. R. 1989. Sources of nitrate to ground water. In *Nitrogen management and ground water protection,* ed. R. F. Follett, 23–33. Amsterdam: Elsevier.

Kirschenmann, Fred. 1988. *Switching to a sustainable system: Strategies for converting from conventional/chemical to sustainable/organic farming systems.* Windsor, N. Dak.: Northern Plains Sustainable Agriculture Society.

———. 1992. Eradicating field bindweed. *Synergy* 4 (1): 17.

Klopfenstein, Terry. 1991. Low-input, high-forage beef production. In *Sustainable agriculture research and education in the field,* 266–78. Washington, D.C.: National Academy Press.

Knutson, Ronald D., Robert C. Taylor, John B. Penson, and Edward G. Smith. 1990a. *Economic impacts of reduced chemical use.* College Station, Tex.: Knutson and Associates.

———. 1990b. Economic impacts of reduced chemical use. *Choices,* Fourth Quarter, 25, 27, 29, 31.

Kuck, Mike, and Jeanne Goodman, eds. 1991. *Ten-year report: Oakwood Lakes–Poinsett Rural Clean Water Program.* N.p.: n.p.

Lamond, Ray E., David A. Whitney, Larry C. Bonczkowski, and John S. Hickman. 1988. *Using legumes in crop rotations.* Cooperative Extension Service pub. no. L-778. Manhattan: Kansas State University.

Langer, D. K, and G. W. Randall. 1981. Corn production as influenced by previous crop and N rate. *Agronomy Abstracts* 182.

Lengnick, Laura L., and Larry D. King. 1986. Comparison of the phosphorus status of soils managed organically and conventionally. *American Journal of Alternative Agriculture* 1: 108–14.

Lewis, James A. 1978. *Landownership in the United States.* U.S. Department of Agriculture Information Bulletin no.435.

Logsdon, Gene. 1986. The *New Farm* visits the *Farm Journal. New Farm,* (May–June): 8–11+.

Lowrance, William W. 1976. *Of acceptable risk: Science and the determination of safety.* Los Altos, Calif.: William Kaufman.

Madden, J. Patrick. 1990. The economics of sustainable low-input farming systems. In *Sustainable agriculture in temperate zones,* eds. Charles A. Francis, Cornelia B. Flora, and Larry D. King, 315–41. New York: John Wiley and Sons.

Marten, John F. 1990. Comments on alternative agriculture: A report prepared by the National Research Council. In *Alternative agriculture scientists' re-*

view, 111–20. Council for Agricultural Science and Technology Special Publication no.16. Ames, Ia.

Martin, John H., Warren H. Leonard, and David L. Stamp. 1976. *Principles of field crop production.* 3d ed. New York: Macmillan.

Mattingly, John C. 1985. This 'feed' kills weeds. *The New Farm,* Sept.–Oct., 35–37.

Midwest Plan Service. 1987. *Beef housing and equipment handbook.* 4th ed. Ames, Ia.: Midwest Plan Service.

Miller, Raymond, and Roy L. Donahue. 1990. *Soils: An introduction to soils and plant growth.* 6th ed. Englewood Cliffs, N.J.: Prentice Hall.

Monsanto. 1992. Encapsulation grower videotape 114-91-11. Schwartz & Associates Productions.

National Research Council. 1987. *Regulating pesticides in food: The Delaney paradox.* Washington, D.C.: National Academy Press.

———. 1989. *Alternative agriculture.* Washington, D.C.: National Academy Press.

Nebraska Ag Report. 1990. Oct.

Neumann, A. L. 1977. *Beef cattle.* 7th ed. New York: John Wiley and Sons.

Olson, Kent D., James Langley, and Earl O. Heady. 1982. Widespread adoption of organic farming practices: Estimated impacts on U.S. Agriculture. *Journal of Soil and Water Conservation* 37: 41–45.

Osteen, Craig D., and Phillip I. Szmedra. 1989. *Agricultural pesticide use trends and policy issues.* U.S. Department of Agriculture Economic Research Service Report no.622.

Peters, Steve. 1991. Ten years without herbicides. *The New Farm.* March–April, 9.

Pfeiffer, T. W. 1987. Selection for late-planted soybean yield in full-season and late-planted environments. *Crop Science* 27: 963–67.

Pimental, David, ed. 1991. *Handbook of pest management in agriculture.* 2d ed. Vol.1. Boca Raton, Fla.: CRC Press.

Potash and Phosphate Institute. n.d. "Livestock manure—why it's a limited resource for profitable crop production systems." Pamphlet. Atlanta, Georgia.

Rachman, Nancy. 1989. Risk assessment: An imperfect science. *Farm Chemicals,* Winter, 22–24.

Raymor, P. L., and R. L. Bernard. 1988. Response of current soybean cultivars to late planting. *Crop Science* 28: 761–64.

Reganold, John P. 1988. Comparison of soil properties as influenced by organic and conventional farming systems. *American Journal of Alternative Agriculture*. 3: 144–55.

Rescher, Nicholas. 1983. *Risk: A philosophical introduction to the theory of risk evaluation and management*. Lanham, Md.: University Press of America.

Robbins, John. 1987. *Diet for a new America*. Walpole, N.H.: Stillpoint Publishing.

Robinson, Ann Y. 1990. *Sustainable agriculture: A brighter outlook for fish and wildlife*. Arlington, Va.: Izaak Walton League of America.

Rodricks, Joseph, and Michael R. Taylor. 1990. Application of risk assessment to food safety decision making. In *Readings in risk,* ed. Theodore S. Glickman and Michael Gough, 143–53. Washington, D.C.: Resources for the Future.

Rosenthal, Sara S. 1983. Field bindweed in California: Extent and cost of infestation. *California Agriculture* 37 (9 & 10), 16–17.

Rosenthal, Sara S., Lloyd A. Andres, and Carl B. Huffaker. 1983. Field bindweed in California: The outlook for biological control. *California Agriculture* 37 (9 & 10), 18.

Sander, D. H., and K. D. Frank. 1980. Fertilizing grain sorghum. *NebGuide* no. G74-112. Lincoln, Nebr.: Cooperative Extension Service, Univ. of Nebraska.

Saxton, Keith E., Ralph G. Spomer, and Larry A. Kramer. 1971. Hydrology and erosion of loessial watersheds. *Journal of the Hydraulics Division, Proceedings of the American Society of Civil Engineers* 97 (November): 1835–51.

Schaller, Neill. 1991. *An agenda for research on the impacts of sustainable agriculture*. Greenbelt, Md.: Institute for Alternative Agriculture.

Schauer, Anne, ed. 1991. *The Thompson farm on-farm research*. Emmaus, Pa.: Rodale Institute.

Schneider, Keith. 1991. U.S. seeks to learn if facts on pesticides were falsified. *New York Times,* 2 March, 8.

Senft, Dennis. 1990. Mighty mite takes on bindweed. *Agricultural Research,* Oct., 26.

Shrader-Frechette, Kristin S. 1985. *Risk analysis and scientific method*. Dordrecht: Reidel.

Singer, Peter. 1990. *Animal liberation*. 2d ed. New York: New York Review of Books.

Smith, Darrell. 1992. No quick fix for compaction. *Farm Journal,* March, AC-5, AC-8.

Spomer, Ralph. G., Keith. E. Saxton, and H. G. Heinemann. 1973. Water yield and erosion response to land management. *Journal of Soil and Water Conservation* 28: 168–71.

Strange, Marty. 1988. *Family farming*. Lincoln: Univ. of Nebraska Press; San Francisco: Institute for Food and Development Policy.

Sutter, Gerald R. 1991. New strategies for reducing insecticide use in the Corn Belt. In *Sustainable agriculture research and education in the field,* 231–49. Washington, D.C.: National Academy Press.

Thomas, Fred. 1990. Midlands rain carries herbicide traces. *Omaha World Herald,* 23 April, 9, 11.

Thompson, Richard, and Sharon Thompson. 1991. 1991 field day handout sheets. Boone, Ia.: Thompson On-farm Research.

Troyat, Henri. 1986. *Chekhov*. Translated from French by Michael H. Heim. New York: Dutton.

U.S. Department of Agriculture. 1980. *Report and recommendations on organic farming*. Washington, D.C.: U.S. Government Printing Office.

———. 1991. *Agricultural chemical usage 1990 field crops summary*. Washington, D.C.: Economic Research Service.

U.S. Department of Commerce. 1987. *1987 census of agriculture*. AC87-A-51. Washington, D.C.: Bureau of the Census.

U.S. Soil Conservation Service. 1982. *National resources inventory*. Washington, D.C.: U.S. Dept. of Agriculture.

———. 1989. *1982–87 Natural resources inventory: Nebraska*. Washington, D.C.: U.S. Dept. of Agriculture.

———. 1992. *Crop residue management guide*. Publication SCS-CRM-01. Washington, D.C.: U.S. Dept. of Agriculture.

Van Arsdall, Roy, and Kenneth E. Nelson. 1983. *Characteristics of farmer cattle feeding*. Agriculture Economic Report Number 503. Washington, D.C.: U.S. Dept. of Agriculture.

Warnock, John W. 1987. *The politics of hunger*. New York: Methuen.

Watson, Maurice E. 1990. Alternative agriculture: Have we been there before? In *Alternative agriculture scientists' review,* 92–95. Council for Agricultural Science and Technology Special Publication no.16. Ames, Ia.

Williams, Martin W., Patrick W. Holden, Douglas W. Parsons, and Matthew N. Lorber. 1988. *Pesticides and ground water data base: 1988 interim report.* Washington, D.C.: Environmental Protection Agency, Office of Pesticide Programs.

Working Group on Synergy in Complex Mixtures, Harvard School of Public Health. 1986. Synergy: Positive interaction among chemicals in mixtures. *The Journal of Pesticide Reform* 6 (2): 11–14.

World Bank. 1986. *Poverty and hunger: Issues and options for food security in developing countries.* Washington, D.C.: World Bank.

World Commission on Environment and Development. 1987. *Our common future.* Oxford: Oxford Univ. Press.

World Health Organization. 1990. *Public health impact of pesticides used in agriculture.* Geneva: World Health Organization.

Zahradnik, Fred. 1984. Graze low-cost pastures through January. *New Farm,* July–Aug., 32–33.

• • •

Index

Other volumes in the series Our Sustainable Future:

Volume 1
Ogallala: Water for a Dry Land
By John Opie

Volume 2
Building Soils for Better Crops: Organic Matter Management
By Fred Magdoff

Volume 3
Agricultural Research Alternatives
By William Lockeretz and Molly D. Anderson

Volume 4
Crop Improvement for Sustainable Agriculture
Edited by M. Brett Callaway and Charles A. Francis